SOPHIE GERMAIN

STUDIES IN THE HISTORY
OF MODERN SCIENCE

Editors:

ROBERT S. COHEN, *Boston University*

ERWIN N. HIEBERT, *Harvard University*

EVERETT I. MENDELSOHN, *Harvard University*

VOLUME 6

LOUIS L. BUCCIARELLI and NANCY DWORSKY

Massachusetts Institute of Technology, Cambridge, Massachusetts

SOPHIE GERMAIN

An Essay in
the History of the Theory of Elasticity

D. REIDEL PUBLISHING COMPANY

DORDRECHT : HOLLAND / BOSTON : U.S.A.
LONDON / ENGLAND

Library of Congress Cataloging in Publication Data

Bucciarelli, Louis L
    Sophie Germain : an essay in the history of the
theory of elasticity.

    (Studies in the history of modern science ; v. 6)
    Includes bibliographical references.
    1. Elasticity—History. 2. Germain, Sophie, 1776–
1831. I. Dworsky, Nancy, joint author. II. Title.
III. Series.
QA931.B77   531'.3823'01   80–19606
ISBN 90–277–1134–8
ISBN 90–277–1135–6 (pbk.)

Published by D. Reidel Publishing Company,
P.O. Box 17, 3300 AA Dordrecht, Holland

Sold and distributed in the U.S.A. and Canada
by Kluwer Boston Inc.,
190 Old Derby Street, Hingham, MA 02043, U.S.A.

In all other countries, sold and distributed
by Kluwer Academic Publishers Group,
P.O. Box 322, 3300 AH Dordrecht, Holland

D. Reidel Publishing Company is a member of the Kluwer Group

Printed in The Netherlands

for Ann

PREFACE

Why should the story of a woman's role in the development of a scientific theory be written? Is it to celebrate, as some have done, the heroism of a woman's struggle in a man's world? Or is it, rather, to demonstrate that gender is irrelevant to the march of scientific ideas? This book hopes to do neither. Rather, it intends to do justice both to the professional life of a woman in science and to the development of the theory with which she was engaged.

Technically, this essay centers on Sophie Germain's analysis of the modes of vibration of elastic surfaces, work which won a competition set by the French Academy of Sciences in 1809. It also evaluates related work on the mathematical theory of elasticity done by men of the Academy. Biographically, it is about a woman who believed in the greatness of science and strove, with some measure of success, to participate in that noble, but wholly male-dominated, enterprise. It explores her failures, analyzes her success, and describes how the members of the Parisian scientific community dealt with her offerings, contributions and demands.

Laying the groundwork for the interaction between Sophie Germain and the scientific world of Napoleonic France has required a chapter on the evolution of Laplace's molecular way of modeling all physical phenomena, showing how Laplace's ideas influenced the emergence of elasticity theory. In this respect, our essay complements the work of other historians: Frankel's 'Corpuscular Optics and the Wave Theory of Light'; Fox's 'The Rise and Fall of Laplacian Physics', and Crosland's book *The Society of Arcueil*.

This study should be of interest also to scholars and students of history who have little facility in mathematics and physics. In writing every chapter we have tried to summarize and evaluate technical matters in less specialized language. Although the substance of the letters, memoirs and events deals with matters scientific, their meaning and motivation transcends the bounds of scientific proficiency and is worthy of a more humane reading and interpretation.

LIST OF ILLUSTRATIONS

TABLE OF CONTENTS

ACKNOWLEDGMENTS

Over the past ten years, many individuals have, in many different ways, contributed to this endeavor. A special word of thanks is due to Nathan Sivin for his constant support and encouragement over the whole duration of this project. To Duncan Foley we owe our gratitude for help in making a translation of the Latin ode by Villoison. We are indebted to Irving Kaplan, for assistance in translation of German sources, for critical review of our manuscript at various stages in its development, and for the pleasure of frequent informal discussion with a humane scholar and professor of nuclear engineering.

The following libraries and archives allowed easy access to source materials and have granted us permission to quote, in some instances rather extensively, from materials in their collections: Archives de l'Académie des Sciences de Paris, Bibliothèque de l'Institut de France, Bibliothèque nationale, the library of the Ecole Nationale des ponts et chausées, the central library of the Ecole Polytechnique, and the Universitätsbibliòthek, Göttingen. Responsibility for translations from the French rests with L. Bucciarelli.

This work was supported in part by a grant from the National Science Foundation.

INTRODUCTION

In 1809, the First Class of the Institute of France, the Science Academy's section devoted to the pursuit of excellence in mathematics and physics, established a *prix extraordinaire*, a medal of gold valued at 3000 francs to be awarded to the person who could provide an analysis explaining a new and striking physical phenomenon: the modes of vibration of thin, flat, elastic plates. Lagrange, Biot, Laplace, and Legendre, as members of the First Class, could not enter the competition, although each had his own view on how to attack the problem. Fourier, Navier, Cauchy, and Poisson[1] (especially Poisson) might have attempted a solution, but none of these aspiring *savants* entered the competition. The prize was won by Sophie Germain.

The present study focuses on that contest and on the woman who was awarded the prize as a way of exploring early developments in the mathematical theory of elasticity. Today, that theory defines not only the modes of vibration of plates, but also how all solid elastic bodies will behave when they are subjected to either static or dynamic loading, a behavior measured in terms of deformation, vibration, deflection, and failure. Although most actual engineering problems do not require working with the general theory itself but are soluble within the framework of specialized analysis, all structures from bridges to aircraft wings conform to it. One theory, based on certain principles and certain assumptions that define the conditions under which the mathematics is operable, accounts for all cases. This theory also implies a conceptual framework through which matter and the forces acting on and within it are understood.

The elegance of this modern theory and the impressive breadth of its applications provided the initial motivation for this historical study. When we began, the *prix extraordinaire* was nowhere within view. We wanted to discover where this theory had come from and how it developed. Such questions led, reasonably, to a traditional search for significant discoveries and the men responsible for them. With Todhunter and Pearson's three volume *History of the Theory of Elasticity*[2] in hand, the search proved easy. Shaping a coherent evolution based on the varied selections contained in this sourcebook, however,

1

proved more difficult. It became a task akin to reconstructing a family history from an album of photographs chronologically arranged, taken from many different perspectives, in various locations, but only on festive occasions. Todhunter and Pearson had strung together discrete and disjointed events into a series in which chronology served as a substitute for causation. Their vision of a theory spawned by Leonard Euler, nurtured by Navier and Poisson, brought to maturity by Cauchy, and to perfection by a host of their successors appeared a ragged and unconvincing march of progress.

Seeking a more cogent and convincing history, we turned our attention to the original memoirs written by these scientists, first singling out for study Navier's 'Mémoire sur les lois de l'equilibre et du movement des corps solides élastiques.'[3] Other historians, including Todhunter and Pearson, had said that this particular treatise was crucial. Immediately we confronted the challenge of creating history, for the way in which this first form of a mathematical-physical theory could have emerged from the chaos of ideas, techniques, observed phenomena, and related theories that preceded it, was totally obscure. It would be necessary to find some perspective that would render these elements essential and harmonious prerequisites to Navier's work. His treatise was disconcerting on another ground: he relied on the same physical principles that the modern theory of elasticity uses, and on much the same assumptions; but his conceptual framework, which, today, we would call his model, differed from the modern one. Today, the theory of elasticity considers an elastic medium as a continuum, a continuous mass of matter through which stress and strain can be distributed. Navier worked with neither stress nor strain – those terms became meaningful only within the conceptual framework of matter as a continuum. Instead, he worked with what is known as a corpuscular hypothesis, i.e. he saw all matter composed of discrete particles, with forces coming into play between them as the distances between them changed. Hence Navier dealt with displacement of points and the forces between them, rather than with stress and strain.[4]

Navier's theory was clearly derived and the form of the equations he obtained relating the displacements of points within an elastic body to the forces acting on that body was correct, but the way in which the specific properties of different materials entered into these equations was deficient. The inadequacies of this first theory (errors according to today's model) appeared to lie in his use of a molecular, or corpuscular, force scheme as a basis for analysis.[5]

Intrigued by his model, we put aside subsequent significant developments in theory (Cauchy's memoirs on the subject were next on our list) and turned to an exploration of what we have since chosen to call a molecular mentality. Our study shifted to a reconstruction of the intellectual milieu that sustained Navier and within which he and others worked.

Navier was not alone in his use of a molecular model as a basis for the analysis of physical phenomena. Pierre Simon de Laplace, France's intellectual heir to Newton, and a most important figure in the history of science, had developed and actively promoted the molecular conceptual scheme as a basis for explaining terrestrial (as opposed to celestial) phenomena. He had claimed that the behaviour of light, the flow of heat, electricity and magnetism, fluid statics incuding capillary action, chemical affinities, the physics of gases, as well as the deformation of elastic bodies, were all subject to molecular analysis. During the first decade of the nineteenth century this way of modeling appeared to possess the potential for explaining all phenomena.

In such an intellectual climate it was not surprising to find Navier working within the molecular hypothesis. The same is true for Poisson and Cauchy who were also exploring elasticity theory. It would have been more surprising had they not been using this model. But nothing obvious could explain the acrimony of their disagreements on fine points, or justify an especially vitriolic polemic between Poisson and Navier, published in a prestigious scientific journal of the time.[6]

Thus our preliminary foraging among original source materials raised new kinds of questions at the same time that it brought a measure of coherence to some of the elements of Todhunter and Pearson's mosaic of events. These questions concerned the way scientists, the people within a profession, dealt with ideas and with one another and with outsiders whose opinions and actions influenced their research. Apparently ingenuity of theory, rigor of analysis, and clarity of experimental evidence were not the only determinants of progress in the sciences. Rather science, like every other endeavor, was rooted in human lives and in the interplay between people and institutions. The hegemony of the molecular mentality at this time had as much connection to Laplace's position of power and authority within the scientific community as to its utility in explaining diverse phenomena.

Moving back from Navier's 1821 memoir, we located the juncture where the molecular mentality first made contact with the elastic behavior of solids and discovered that Laplace had indeed been very much involved in events at that time. The stimulus for all the contemporary activity in elasticity lay in the experimental demonstration of the vibration modes of plates to the members of the First Class of the Institute in the fall of 1808. That dramatic yet simple experiment was performed by a German scholar, acoustician, and correspondent of the *Societé Philomatique de Paris*: E. F. F. Chladni.

Chladni had come to Paris because Paris was the center of scientific excellence in the western world, a place where the study of science was publicly acknowledged to be a high and important human endeavor honored by state support and encouragement. At the time of his visit, Paris was the center for a golden era in the development of mathematics; a time when mathematics matured as a distinct area of inquiry and a tool for explaining natural phenomena. Although this process had begun before the French Revolution, and continued without rupture or diversion through the Napoleonic era, Napoleon's patronage of science and mathematics – a practice inherited from the eighteenth century and suited to his own ambitions – must account, to a degree, for the advancement of science in France at that time.[7] Evidence of the high quality of work in the sciences during the early decades of the nineteenth century in Paris is revealed by the list of names to be found in the historical footnotes of modern science and engineering texts; labels that identify a variety of special solutions, differential equations, and techniques used in twentieth-century mathematics: Fourier transforms, Legendre polynomials, Laplace's equation.

Some of these men were already established scientists in 1808, setting the standards by which others, aspiring to recognition, would be judged. As members of the First Class of the Institute, the class of mathematical and physical sciences, they formed an elite group of sixty, distributed among ten subsections – agriculture, physics, mathematics, astronomy, . . . , zoology.

The First Class was not simply an honorary society; it was the official center for scientific exchange. It met every Monday throughout the year to hear papers by members and outsiders, to establish state-sponsored commissions to assess proposed technical endeavors, to set up competitions on outstanding scientific problems, to review progress in the sciences at home and abroad, and to plan, support, and carry out scientific expeditions.[8] A busy place, a place of business as much as of science, it was the institutional setting, the formal framework, for the workings of a scientific community. And because it was all these things it became the setting for Chladni's demonstration of the modes of vibration of thin, flat elastic plates.

The resonant vibration of elastic bodies had always been an intriguing phenomenon. The Pythagorean cult's near-mystic concern with the resonant mode shapes and frequencies of a vibrating string is well known. An explanation of this singular behavior of a taut cord, an explanation in terms of mathematical analysis, had been provided by d'Alembert, Lagrange, and Euler in the eighteenth century[9] The latter two also had succeeded in generalizing the analysis of a string to include the vibrations of a membrane.

What Chladni showed was that, just as the modes of a resonant string or a beam possess nodes where there is no motion, so a flat plate when caused to vibrate displays curves or lines where there is no motion. These form patterns on a resonating surface that are as unique as the mode shapes of the vibrating beam, and as infinite in their possible design. How Chladni accomplished this display is related in this excerpt from a contemporary review of his book, *Acoustics*:[10]

M. Chladni has discovered a means as simple as it is ingenious for making these curves visible. He first covers the plate with powder. When the plate is excited, the powder abandons all oscillating regions of the body, collects and settles down motionless at the boundaries of these regions. The curved lines of equilibrium so formed take on very different but regular patterns.

To do the experiment it is necessary to grasp the plate with two fingers, the tips squeezing it at two points on opposite edges, and to stroke the plate with a bow at a point on its perimeter. One sometimes applies a third finger to the different points on one of the edges in order to vary the results of the experiment. One can, instead of holding the plate between the fingers, place one of its faces at a fixed point and support the other face at a second point placed exactly opposite the first. . . .

Figure I.1 indicates the enthusiasm with which Chladni applied himself: square, rectangular, polygonic, as well as circular plates yielded a profusion of patterns and tones.[11] The complexity of results is striking in contrast to the simplicity in both concept and materials of the experiment. Yet it is not a simple matter to reproduce Chladni's results. The difficulty arises from the need to hold the plate perfectly level and to stroke correctly with the bow in order to produce a pure tone: only a pure tone will produce a unique mode pattern. The discovery, therefore, of this enormous variety of patterns was not a simple demonstration and caused something of a sensation in Parisian scientific circles.[12]

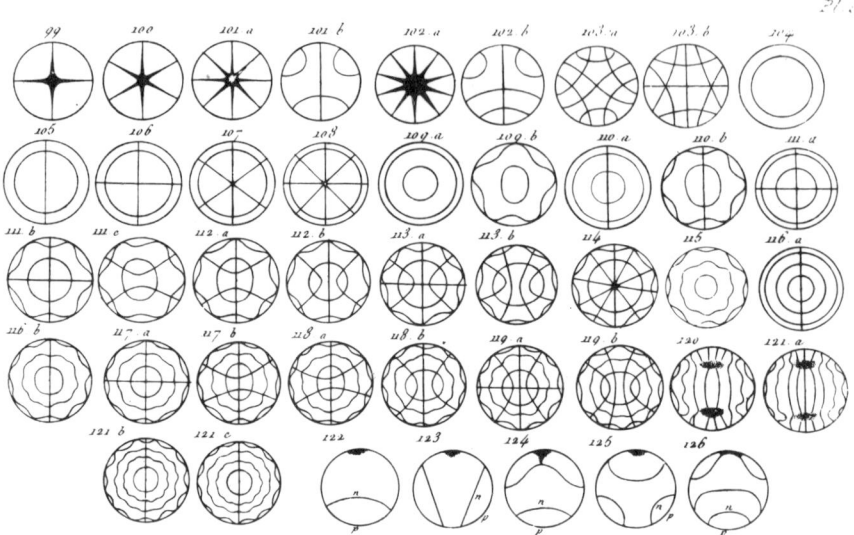

Fig. I.1   Chladni's vibration modes of circular plates.

Indeed Laplace presented Chladni, and Chladni his experiments, to Napoleon. (Laplace had at one time been Napoleon's Minister of the Interior.) Napoleon, in turn, being "struck by the impact which the discovery of a rigorous theory capable of explaining all the phenomena revealed by these experiments would have on the advancement of physics and analysis,"[13] urged the First Class of the Institute to set the problem as the subject of an extraordinary prize. The First Class did not need so formal a stimulus to action. Not only were the experimental phenomena worthy of attention in their own right, but it appears that Laplace had a personal interest in establishing a competition in the area of mathematical physics as a way of providing his *protégé* Poisson, not yet a member of the First Class, with the opportunity to display his talents.

The prize was not easily won; in fact the competition was twice reset. Nor did Poisson win the prize. Sophie Germain, in absentia though with some fanfare, was awarded the 3000 francs at a public session of the Institute in January, 1816. That a woman appeared to have made a significant contribution in the field of mathematical physics aroused our curiosity. Preliminary research revealed that she was self-educated, spent her entire life (1776–1831) in Paris, corresponded with Gauss,[14] was tutored by Legendre and Fourier, and worked on problems in number theory as well as on the vibration of plates.

Turning to her prize-winning work on the latter problem – her analysis was published as a memoir in 1821 – we sought a rigorous derivation of the partial differential equation that defines the vibration of flat, thin elastic plates. The plate equation, her final result, met our expectations, but her analysis was inadequate, and must have appeared so to her contemporaries who awarded the prize. Her deficiencies were due, in part, to a lack of sophistication in arguing her basic hypothesis, in part to neglect in establishing the consistency of an approximation made in the course of her derivation, and in part – most surprisingly – to failure in certain basic technical skills. Her grammar, so to speak, was deficient; she could not spell. "The lady does not appear to have paid the attention to the calculus of variations that might have been expected from the pupil of its great inventor Lagrange."[15]

How had Sophie Germain managed to obtain what was soon recognized to be the correct equation defining the vibration modes of plates from such a derivation? Clearly there are facts here which demand explanation. Why was she awarded the prize on the basis of such flawed work?

Sophie Germain became the focal point in our investigation of early developments in the theory of elasticity. Our interest in her lies at once in her scientific accomplishments, the influences that shaped her work, and the importance of her work to others. Her attempted explanation of Chladni's patterns, although awkward and clumsy when viewed against the background of available mathematics of the time, did lead to a correct equation for the elastic behavior of plates and stimulated Poisson, Navier, and eventually Cauchy[16] to address the behavior of elastic bodies in a general way. But just as the end point of her analysis is of interest, so too is her starting point. Sophie Germain adopted, as a basic hypothesis, an expression for the elasticity of the plate that she claimed did not depend on any special underlying physical mechanism. Others, notably Poisson, devoted considerable energy to framing their theories atop a corpuscular-force model, a way of viewing physical phenomena that, while very much in vogue in 1808, within a few decades following Sophie Germain's receipt of the prize, was relegated to a more humble place in the conceptual baggage of normal scientific practice. Her work, again despite its flaws, became one (of several) rallying points for criticism of that molecular mentality.

While the history of science and the development of elasticity theory is central to our treatise, it would be specious to ignore the other aspect of this study; it has a place in the history of women in this male-dominated culture. From this perspective it becomes a test case of women in science, in particular of eager and talented women, who, had they been men, would have fared far differently. For all the evidence argues that Sophie Germain had a mathematical brilliance that never reached fruition due to a lack of rigorous training available only to men. Furthermore, Sophie Germain's experience clarifies a whole structure of social expectation and behavior – no less powerful for being unconscious – that has kept women from full professional activity. This aspect of our study lies outside the history of elasticity theory, but is addressed here as an interesting, if timely and somewhat problematic, ingredient of a more socially-oriented and more meaningful, history of science.

CHAPTER TWO

SOPHIE GERMAIN

Marie-Sophie Germain was born on the first of April, 1776, into a Parisian bourgeois family that had attained a comfortable level of prosperity through several generations of trade.[1] Her father, Ambroise-Francois Germain, was a silk merchant who became briefly but actively engaged in politics at the beginning of the French Revolution as an elected deputy of the third estate to the Constituent Assembly that convened in 1789. He died in 1821 at the age of ninety-five.[2] Sophie, the second of three daughters, remained financially dependent upon her father throughout her life. Never marrying, she devoted her life to the study of mathematics and science.

Her sisters led more conventional lives. The eldest, Marie-Madeleine, born 20 May, 1770, married a notary, Charles Lherbette. Their one child, Jacques-Amant, born 16 September, 1791, led an active life as a lawyer, sportsman, and politician. He was clearly a favorite of his aunt Sophie.[3] Sophie's younger sister, Angelique-Ambroise, born 1779, married into medicine. Her first husband, Rene-Claude Geoffroy, a doctor, died in 1831. She remarried in 1833 the physiologist, physician, and member of the Academy of Sciences, Joachim-Henry Dutrochet. He died in 1847, but Madame Dutrochet went on to live another quarter century, attaining like her father, the age of ninety-five.

Angelique-Ambroise's first marriage brought an improvement in the family's living accommodations. In 1776, the year Sophie was born, they lived atop the family shop located near the intersection of Rue St. Denis and Rue des Lombards, close by the Marche des Innocents. Prior to 1798 the family moved a few blocks to 23 Rue Sainte-Croix de la Bretonnerie. Then, some time shortly after 1816, Sophie and her parents moved from these relatively modest quarters to join the Geoffroys at their grand town house, which still stands at 4 Rue du Braque, a short distance from the National Archives.[4]

During Sophie's childhood, however, the Germains' fortunes had not been so secure. The political ferment of the French Revolution was part of family life and may indeed have been significant for her future career in mathematics and science. As a thirteen-year-old girl, shy, rather awkward, and personally timid, she found her home filled with talk of money, politics, and change. She went into her father's library to lose herself in books, to discover a world that was beautiful, exciting, and promising but threatening. In particular, she found two books that captured her attention: Bezout's standard text in mathematics, a book liable to be found at that time on every educated person's bookshelves, and Montucla's *Histoire des Mathématiques*. In Montucla, she read of the heroic accomplishments of Archimedes: "Such was his passion for these sciences [Mathematics and Mechanics] that he would forget food and drink. His servants would have to remember them for him and would almost have to force him to satisfy these human needs."[5] Not even the uproar and confusion of the Roman invasion of Syracuse could distract him from his favorite studies.

It is neither outlandish nor sentimental to imagine a thirteen-year-old searching for assurance in a terrifying world, impressed enough by this account to seek in mathematical studies an environment where she too could live untouched by the confusions of social reality. This youthful withdrawal set a pattern for her entire life.

Certainly Sophie Germain approached mathematical studies with limitless passion and devotion. G. Libri relates in an obituary article how she surmounted "all obstacles with which her family first tried to impede so extraordinary a taste for her age, no less than for her sex, by getting up at night in a room so cold that the ink often froze in its well, working enveloped with covers by the light of a lamp even when, in order to force her to rest, her parents had put out the fire and removed her clothes and a candle from the room."[6] This may be a somewhat embellished account of household realities, but there is no reason to suspect that Libri invented it wholly to glorify his subject at the expense of her family, or that Sophie Germain recounted such tales with no foundation in fact. Nor do we have reason to doubt that the family could have reacted so harshly to such a peculiar predilection in one of its daughters – this strange child in a bourgeois Parisian home, amid the ferment and excitement of revolutionary France, withdrawing into equations and symbols largely foreign to her sex and class, finding her pleasures neither in religion, nor in society, but in coming to understand 'the language of analysis' and, later, the differential calculus of Cousin.[7]

Germain, Sophie (1776–1831), French Mathematician, . . . born in Paris April 1, 1776, took correspondence courses from *L'École Polytechnique*, which did not accept women in the school itself.[8]

With this grand anachronism the *Encyclopaedia Britannica* solves what must have been a difficult problem for Sophie Germain. She was in her late teens in 1795, when the Ecole Polytechnique was established. By that time her study of mathematics had grown from the childish enthusiasm of a thirteen-year-old to the mature and serious study of a determined, talented young woman. Libri writes:

> With the establishment of the Ecole Normale and the Ecole Polytechnique, she obtained for her own use the lecture notes of different professors. Fourcroy's chemistry and Lagrange's analysis especially drew her attention. At that time, these professors, at the end of their courses, employed the excellent practice of requiring their students to present their observations in writing. Mademoiselle using the name of a student at the Ecole Polytechnique, sent hers to Lagrange.[9]

The name she used was LeBlanc. Registered among the first year's class of the Ecole Polytechnique was a student by the name of Antoine-August LeBlanc. A year older than Sophie Germain, he had grown up in Paris and upon graduation was admitted to the Ecole des Ponts et Chausées,[10] an excellent school, whose curriculum was oriented towards practical engineering pursuits and which required rigorous training in mathematics and science for entrance. How he came to meet Sophie Germain remains unknown. Without wanting to slander the unfortunate Antoine, who died at the age of twenty-two before he could attend the Ecole des Ponts et Chausées, it is tempting to imagine a friendly arrangement whereby Antoine attended the classes and Sophie Germain did the homework.[11] Is this as unlikely an anachronism as the *Britannica's* correspondence course? In fact, there is no way of knowing what the social mores of the time allowed. We know that Sophie Germain was somewhat of a recluse and avoided social engagements; we also know that she was far from shy where learning was at stake. But how much real possibility and flexibility were open to a woman of her class with her intellectual ability and interests, we do not know: social history does not normally concern itself with such things.

In any case, it was through the Ecole Polytechnique that Sophie Germain came in contact with Lagrange. When she responded to his lectures under the name of LeBlanc, Lagrange, impressed by her work, "then learned the true name of its author and went to her to express his astonishment in the most flattering of terms."[12] Exactly how Lagrange learned of her identity is unrecorded, as is the response of Sophie Germain or her family to his discovery. One is tempted to conjure up scenes of an impressed professor seeking out a brilliant student. Embarrassed and bumbling, LeBlanc admits the paper was written not by him, but by his friend Sophie Germain.

Though lacking specific information about encounters, we do have Libri's comment that "the appearance of this young 'geomètre' made quite a stir and Mademoiselle Germain did not have to wait long to see scientists of superior merit coming to her; their conversations provided nourishment for her mind." A few letters have been preserved from this period that give substance to Libri's statement.

Among them we find one from a Monsieur Bernard addressed to Sophie Germain's mother on behalf of the aging 'Citizen' Cousin. The author of the popular *Leçons sur le calcul différentiel et le calcul intégral* requested a meeting with Mademoiselle Germain and offered to place at her disposal all the resources he possessed for the practice and profession of science.[13] Other *savants* who took an interest in Sophie Germain seem to have been just as eager to aid in her education, however the quality of problems they tried to explain was very uneven. For example, a note that may have been authored by Cousin, or perhaps by Lagrange, addressed several rather trivial mathematical paradoxes: one due to Monge, which involved the equilibrium of a lever (an infinitely small, or zero, force at an infinite distance from a fulcrum can move a finite weight located at a finite distance); another treated a geometrical representation of this same paradox in terms of areas under a hyperbola; and a third was the proposition "that 0/0 is equal to anything one wishes." The author located the source of these absurdities in a faulty conception of the infinitely large: "One falls into these paradoxes every time one transports oneself to the land of the infinite; to state that it is absurd that one could ever get there is not just a figure of speech."[14]

Paradoxes may have been a fashionable topic toward the end of the eighteenth century, but these questions were not inherently worthy of the careful explication that her correspondent provided. His meticulous attention testifies to his willingness to give whatever help he could to Sophie Germain. At the same time, this note supports an impression of haphazardness in her education: books, lecture-notes, problems, came by more as chance would bring them than as orderly curriculum would provide.

Not all of Sophie Germain's exchanges with mathematicians were comfortable and encouraging. A small crisis resulted from a visit of the astronomer Lalande, then sixty-five years old, an active and well-known figure in intellectual society.[15] We know nothing about the substance of the incident, but something in Lalande's behavior aroused Sophie Germain's indignation, as the following letter from him reveals.

au College de France

Mademoiselle   4 Nov. 1797

It would be difficult for anyone to make me feel more the imprudence of my visit and the disapproval of my respects than you did yesterday. It would have been difficult, however, for me to have foreseen the outcome.

I still cannot understand or reconcile what happened with the talents that my friend Cousin told me of. All that is left for me to do then, is to apologize for my imprudence. One learns at every age, and the lessons one learns from a person as likeable and wise as you remain longer than those of others.

You told me that you had read Laplace's *Système du Monde* and that you did not wish to read my short work on astronomy. I said that I thought that you could not understand the one without the other. I assume that it was this suggestion that caused your anger. For this I apologize.

Best regards.   Lalande[16]

Exactly what form Lalande's respects took must be left to the imagination, but his letter is mysterious enough to allow conjecture that Sophie Germain was not simply hypersensitive. Whatever his social indiscretion, it seems to have been mixed with a professional insult. Lalande's book, which he suggested that Sophie Germain read, was written especially for the education of women – a simplified text to introduce them to the excitement of serious study and draw them away from pure frivolity.[17] Sophie Germain would certainly have taken umbrage at the suggestion that she belonged among Lalande's intended audience. His image of her was evidently in error from an intellectual as well as social perspective. He was clearly not delighted with the real woman whom he met.

For her part Sophie Germain seems to have been exceedingly angry rather than embarrassed. She evidently let it be widely known that M. Lalande was not to be included among her acquaintances. In her correspondence we find an invitation to a social affair, a dinner, sent by Tessier, another senior member of the First Class of the Institute affiliated with the Agriculture Section. Tessier did his best to encourage her to accept, tempting her with a menu of fresh butter, potatoes, beets, and cornbreads, assuring her that her father need not worry (Tessier would see to it that she would be safely transported home at any specified hour) and informing her that "M. L[alande] will not be there, since you have not reconciled yourself with him."[18]

Though she successfully avoided him in the drawing room, Sophie Germain appeared with Lalande in print in 1802. She was well enough known in intellectual circles for her name to be invoked in occasional poetry written in praise of the sciences. At least d'Ansse de Villoison, a celebrated Greek scholar, had occasion to write the following letter to Sophie Germain.

Mademoiselle,   [1802]
    In spite of your condemnation of a famous astronomer and his friends, I was not able to restrain myself from rendering homage to the truth, and I am eager to offer you the beginning of a section of a latin verse of my composition that is going to appear shortly in the *Magasin Encyclopédique*. You will see there on page 239 a small measure of the respect I have for you, Mademoiselle, which is due you for so many reasons. I would be only too happy if you would convey to Mademoiselle you sister and accept kindly this acknowledgement of my admiration and respect. I am, Mademoiselle, your very humble and obedient servant.

d'Annse de Villoison

The next letter is addressed to Sophie Germain's Mother.

Madame,   This Monday night at midnight, 12 July, 1802
    I find in reopening the letter with which you honor me and in which you press me to give you my word of honor that your orders and those of Mademoiselle your daughter, will be punctually executed, that I have already destroyed my Greek poem and would like to annihilate even the Latin ones. I will be content in spite of this to admire Mademoiselle your daughter in the most respectful silence, and to look on you as the happiest of mothers and most deserving of envy. I will take the liberty of imploring Mlle. Sophie to accept a copy of the second Paris edition, which is going to appear in a very few days with the addition that I had the honor of communicating to her, and that, I am told, Madame, will be immediately followed by a translation into French by a woman whom I have not yet had the honor of meeting but who is working on it now. I will reproach myself all of my life for having composed this piece that so wounded the excessive modesty of Mademoiselle your daughter. I entreat her, as I do Mademoiselle her sister, to please accept my apologies and my assurances of living and eternal regret. . . .

d'Annse de Villoison[20]

Mademoiselle,   14 July, 1802
    I dare take the liberty of offering you the attached, an example of the new edition of my unfortunate piece, with the corrections and additions that I announced to you. M. Pougens, Mademoiselle, had them inserted in the third number of the third year of his *Biliothèque Française,* before I was able to guess that this acknowledgement of the truth would shock your modesty, which is as extraordinary as your talents. I reiterate my apologies and the expression of my lively and eternal regrets; my word of honor that I will not permit myself to speak of you, Mademoiselle, in anything written. My admiration will be forever in silence and bound by the desire to obtain pardon for an involuntary error or fault, and by the

profound respect that I have for Madame your mother and Mademoiselle your sister. Your very humble and very obedient servant.

d'Annse de Villoison.

P.S. – You confess to me, Mademoiselle, that if you are the only maiden who possesses so superior a knowledge of mathematics, you are also the only one who has known and feared the danger of a Greek poem. In all good conscience, I appeal to Mademoiselle your sister, who is so kind. Could you not be good enough to grant me this favor?[21]

Unfortunately, Villoison seems to have been as good as his word in destroying the Greek poem. Excerpts from the Latin ode follow here in English translation:

Birthday Poem for Lalande, the famous astronomer.
by D. G. d'Annse de Villoison

The stars celebrate this day's rising of a star:
Joyfully Lalande rising now clothes the sky
With unaccustomed light, and embraces the world.
With his birth a new calendar is born;
From this day let the learned mark years and count time.
This is your day, praise it, followers of the Muses.
If the earth refuses to speak that name,
Beloved to the Pieran chorus of lovely women, the sky will resound

Impatient he rises to the clouds with the movement of a bird,
Whence his birth had led; His celestial origin draws him:
It pulls, and fiery energy adds quick wings.
His nephew follows him in flight, and his nephew's wife,
And Burchard likewise goes with equal pace.
Ariadne, by whom skilled Germain's visage is already envied,
Sees and dislike what she sees, yielding her crown.
"What new Epigone enters the starry realm?"
She cries. "Most boldly she tries to enter
Our house, Gods stop her flight;
While you can, rein in this Icarian girl;
For her burning effort will conquer giants.
This ambitious woman already wanders in LaPlace's realm!
And drinks the airy fires with greedy gulps!" . . .

It is impossible to reconstruct with accuracy or assurance what Villoison was trying to achieve with this poem. His verse, in elegant Latin hexameters, is a pastiche of classical images and references; the tone is urbane and rather light. Judging from his first letter, he thought that

Sophie Germain should have been pleased to find herself in such company. (And the company included more than Lalande.)

A curious fact about this poem is that half of it concerns women associated with astronomy. In addition, there is an anonymous 'Cytherian dove' to point the way for a future role in astronomy for women. Of course, the question of women's educability in the sciences was one of contemporary interest, but not especially so in 1802. Lalande had published a book on astronomy for women, *Astronomie des Dames*, in 1785, a second edition of which appeared in 1795. A third edition would appear in 1806, two years after Villoison's poem.

Was Villoison, in writing so much about women, doing so to please Lalande personally? It seems likely, especially since Villoison was having difficulty obtaining a post in his field in Paris, and a chair of ancient and modern Greek was created for him at the College de France shortly before his death in 1805.

Whatever Villoison meant, it did not please Sophie Germain. Whatever his intention, she surely would have taken "Laplace's realm" as a reference specifically to the pretext Lalande had advanced for their quarrel. This is especially likely since she is introduced in the poem not as a colleague with the other ascending astronomers, but as a competitor.

Was it only the association with Lalande that Sophie Germain took exception to? She seems to have been quite withdrawn from the normal social life of the time, living not only apart from the society of male scientists, but apart from the society of educated women as well. She may well have resented being counted with Mme. Lalande[23] and the Duchess of Gotha. Being serious was not congruous with being fashionable. The excessive modesty that Villoison as well as others remark upon can contain several elements: on the one hand, there was a real timidity that made her avoid any normal social encounters and situations; and on the other, a private sense of the superior worth of her own undertakings – superior to those of Lalande (a notorious vulgarizer of science), certainly superior to Villoison's offering to a small circle of *savants*, and probably superior to those of other women in science, whose work was not independent of, but subservient to, the men they associated with. She saw her work as part of Man's march toward Truth and Progress – not as part of a contemporary social scene.

As we look back at Sophie Germain growing into adulthood, it becomes clear that the customary markers in a biography do not help us to grasp her character. Her parents disappear behind formalities:

Fig. II.1   Sophie Germain, a sketch from Stupuy,

Fig. II.2   Sophie Germain, a bust by Z. Astruc.

"Citizen Cousin requests the honor of being presented to you and also to Mademoiselle your daughter, if you condescend to agree." Occasions, events, questions of the day are unrecorded. They are lost to history not simply by chance, but because they really were unimportant to her. Experienced as intrusions into her life, she pushed them out again as soon as possible.

The only images of Sophie Germain that exist appear in Figures II.1 through II.4. The drawing itself (Figure II.1) is far removed from the living woman: it is a sketch drawn from a bust (Figure II.2), which in turn was reconstructed from the death mask (Figure II.3). The coin (Figure II.4) reproduced in *Le Grand Larousse*, looks suspiciously like the death mask with hair added.[24] These pictures, clear yet tentative, serve as a symbol for this brief biography. For Sophie Germain emerges from her history with clarity but without certainty, since records of her life, outside of her work, are scant; the details are few and incomplete. She lived by choice in a world of ideas and abstractions peopled by Montucla and Cousin, by the professorial Lagrange and the ephemeral Leblanc. She did not wish to meet others in the streets or houses of the day, but in the purer realm of ideas outside time, where person was indistinguishable from mind and distinctions depended only on qualities of intellect. Thus we miss in her life the smell of the city streets, the ferment of revolutionary times, or even the harmonies of a pastoral retreat. Rather, her life became subsumed in her work, which in turn became a part of the workings of the scientific community, itself a cultural and historical phenomenon. Sophie Germain, as a person of flesh and blood, disappears.

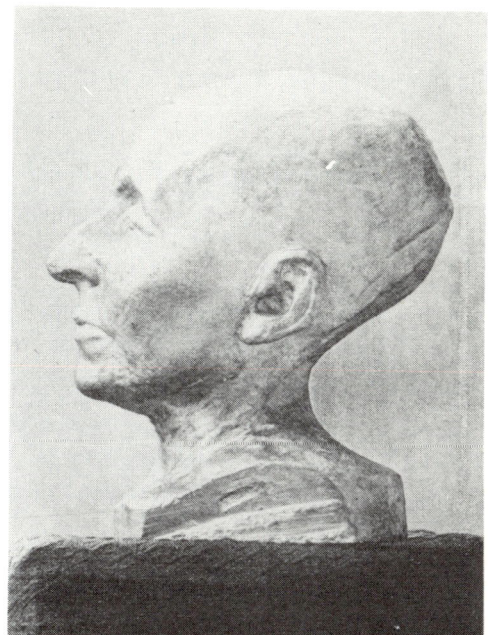

Fig. II.3   Sophie Germain's death mask.

Fig. II.4   A medallion design by Mme. Dufour.

CHAPTER THREE

RESPECTFULLY YOURS, GAUSS

After the publication in 1789 of Legendre's work on the *Théorie des Nombres*, she embarked on a study of this theory with an unwavering enthusiasm. When Gauss' *Disquisitiones Arithmeticae* appeared in 1801, she was amazed by the originality of this famous Göttingen professor's work and experienced another incentive to engage in this kind of analysis. After a number of investigations in this area, she wrote to Gauss, using again the assumed name of a former student of the École Polytechnique. He answered the unknown mathematician in a most respectful manner.[1]

While Sophie Germain was troubled by and impatient with the social attentions of educated men who found her an oddity, she was certainly gratified by the professional recognition of serious scientists. Her major pleasure, however, came not from any satisfaction they gave her ego, but from their instruction which might enrich her studies. If need be, she could escape from undesired social situations by being curt or even rude. It was more difficult to continue serious work and to make a claim for herself as a real mathematician rather than as a favorite of a handful of aging geometers.

As Libri relates, she turned to the theory of numbers, seeking there advancement and accomplishment, no doubt encouraged in this undertaking by Legendre and Lagrange. One can imagine the two of them urging her to test her mathematical abilities against the contents of Gauss's *Disquisitiones Arithmeticae*,[2] a volume rightly praised as a ". . . magnum opus, which in one stroke made number theory a firmly grounded and coherent part of mathematics."[3] She seems to have welcomed this challenge. Number theory, or higher arithmetic (as Gauss preferred) and its attendant array of near-mystical relationships among integers deeply fascinated her and strongly influenced her whole intellectual life.

Within a few years, she had mastered the essential ingredients of Gauss's work and had experienced the joy of discovery as well. Eager for acknowledgment, assurance, and encouragement she wrote to him, adopting again the name of the now deceased Antoine LeBlanc.

In her first letter, dated 21 November, 1804, Monsieur LeBlanc praised Gauss's treatise:

Monsieur,

For a long time your *Disquisitiones Arithmeticae* has been an object of my admiration and study. The last chapter of this book includes, among other remarkable things, the beautiful theorem. . . . Nothing equals the impatience with which I await the sequel to this book I hold in my hands. I have been told that you are working on it at this moment; I would spare nothing in order to procure it as soon as it appears.[4]

Having worked on and completed some exercises, she went on, in a very determined way, to ask for an evaluation of these, her own efforts: "I take the liberty of submitting these attempts to your judgment, persuaded that you would not demur from enlightening with your advice an enthusiastic amateur of the science you cultivate with such brilliant success."[5]

These exercises were diverse in nature; with one exception, all were stimulated by the *Disquisitiones Arithmeticae*. They included two demonstrations of a generalization of Gauss's relationship

$$4 \cdot \frac{(x^n - 1)}{(x - 1)} = Y^2 \pm nZ^2 \text{ to } 4 \cdot \frac{(x^{n^s} - 1)}{(x - 1)} = Y^2 \pm nZ^2$$

(where n is prime and s any number), and an application of this relationship to a problem considered by Lagrange: "Upon rereading Lagrange's memoir (Berlin 1775), I was surprised to see that he was not able to reduce the quantity $s^{10} - 11(s^8 - 4s^6r^2 + 7s^4r^4 - 5s^2r^6 + r^8) \cdot r^2$ to the form $t^2 - 11n^2$ . . . ."[6] A joust with Fermat's last theorem was the one exception alluded to above: "I have added to this letter several other considerations. The last concerns Fermat's celebrated equation $x^n + y^n = z^n$ for which the impossibility of satisfaction with whole numbers has only been demonstrated for $n = 3$ and $n = 4$. I believe it possible to prove this impossibility for $n = p - 1$, p being a prime number of the form $8k + 7$."[7]

Although the details of most of these investigations included as appendices to her letter are no longer available. Gauss's response to Monsieur LeBlanc shows that her efforts were not without merit: "I read with pleasure the things you chose to communicate; it pleases me that arithmetic has acquired in you so able a friend. Your new proof concerning the prime numbers for which 2 is a residue or nonresidue especially pleased me. It is very fine, although it seems to be an isolated case, inapplicable to other numbers."[8]

Encouraged by this response, Sophie Germain sent Gauss other examples of her work.[9]

Monsieur,   21 July, 1805
    It is without a doubt that I owe the flattering response you have seen fit to make to my
letter solely to your indulgence; you give me hope of our continuing this discussion of your
studies; nothing in the world would give me more pleasure than that. Indeed, what a
difference between the feeble attempts of which I am capable and the ingenious methods
whose invention you are intimate with! However, since you have favorably entertained the
notes that I have communicated to you, I take the liberty of sending you some new ones.
You promise (no. 267) to prove, upon another occasion, that the tertiary form whose
determinant is zero is equivalent to a binary form; I have sought to accomplish this reduction
and I have found, in this case, that the adjoint form reduces to a square multiplied by m,
which is excactly the same form that the binary form takes when one assumes its
determinant is zero. . . ."[10]

Despite her solicitousness, Gauss rarely provided the critical reviews she
requested. Except for the restrained praise he gave her new proof
concerning prime numbers for which 2 is a residue or nonresidue and the
presentation of a counterexample negating her proposition concerning
the sum of nth powers of two numbers in the form $h^2 + nf^2$ (see below),
Gauss remarked only on the work she did in relation to his theorems; he
did not become involved with the original work that she sent him. This
was probably less a judgment on the quality of her intellect than a
reflection of his self-absorption.[11]
    Sophie Germain, on her part, was quite willing to bolster his ego. As
LeBlanc she wrote: "Unfortunately the reach of my intellect does not
equal the vivacity of my appetite and I feel a kind of rashness in troubling
a man of genius when I have no other claim to his attention than an
admiration necessarily shared by all his readers."[12] In another place we
find, "I regard as a favor your permission to communicate to you my
feeble attempts, persuaded that you will have the goodness to warn me of
errors in this area of research where you are the lone enlightened judge
that one can consult."[13]
    Sophie Germain, as an "enthusiastic amateur" and his "assiduous
reader," never wavered in her praise for Gauss and responded instantly
to his letters. He, as "the lone enlightened judge" available, felt no such
obligation. "I must beg a thousand pardons for having let six months go
by without answering the kind letter with which you honored me. . . ."[14]
Only on one occasion did Gauss respond promptly and with real
enthusiasm – upon discovering that Monsieur LeBlanc was, in reality, a
woman.
    The circumstances of this discovery were as follows. In 1806

Napoleon's armies were drawn into conflict with Prussian forces. The success of the French Army at Jena in October opened the way for the occupation and siege of the greater part of Prussia. Sophie Germain, recalling the fate of Archimedes in similar circumstances, became concerned about Gauss's safety. She requested a favor of a family friend, M. Pernety, commander of the French artillery in the Prussian campaign, who was responsible for maintaining the siege at Breslau.[15] Sophie Germain charged him with discovering Gauss's whereabouts and ensuring that he not be mistreated. M. Pernety ordered M. Chantal, a battalion commander, to travel some two hundred miles west to the occupied city of Brunswick to carry out this mission. As a result of her intervention into this Napoleonic campaign, LeBlanc was to emerge in her own person.

Chantal's letter to his general, written while he fulfilled his orders on his mission in Brunswick, follows:

General Pernety   Brunswick, 27 Nov., 1806
Chief of State Major-General of the Artillery of the Army,
   Just arrived in this town and have busied myself with your errand. I have asked several persons for the address of Gauss, at whose residence I was to gather some news on your, and Sophie Germain's, behalf. M. Gauss replied that he did not have the honor of knowing you or Mlle. Germain, but that he had known a Madame Lalande in Paris.[16]
   After I had spoken of the different points contained in your orders, he seemed a little confused and asked me to convey his thanks for your consideration on his behalf. If he were to write to Paris, I asked him to give me the letter, then I would give it to you and you would deliver it to its destination. He replied neither yes nor no on this point. I left him there with his wife and child. I visited General Buisson, mayor of the city, in order to recommend [Gauss] to him. I had the honor of meeting the general previously. He replied he would do everything he could for M. Gauss and invited me to dine with him. M. *le commandant*, who was present at that time, told me that this man had already been recommended to him by several persons of merit. I took leave and returned to M. Gauss's residence and asked him to dine with me at the governor's. (Having accepted the invitation, I agreed to come by in an hour so that we would go together.) The result is that he will have from the governor and the commandant of this place all the esteem and kindness at their disposal. On the way to the governor's, I will try to persuade him to write to you about the way in which I conducted my mission at the same time that he writes to Paris, if he thinks it appropriate; I will leave him your address for that purpose. His is: *M. le docteur* Gauss, at Ritter's, Steinweg no. 1917, at Brunswick. He enjoys good health and told me that he was a little frightened when the troops entered Brunswick, but that he was not disturbed. I reassured him and I do not doubt that the commander and the governor will reassure him better in this respect. I have run the post night and day until this moment. This circumstance obliges me to remain here this afternoon; early tomorrow morning I will leave for my destination.

Deign to accept, my general, the sentiments of the most profound respect with which I have the honor to be,

Chantal
Chef de Bataillon[17]

Sophie Germain was informed of these events in the following letter from her friend, General Pernety:

Cotel, near Breslau
Mademoiselle,   23 December, 1806

I cannot make any better reply to the requests that your love for scientists has made of me than by sending you the letter from the artillery officer whom I sent to Brunswick to find out about M. Gauss. I hope that it satisfies your wishes for this rival of Archimedes, better treated than he, as you will see. I hope that this will put me in the position of being able to perform more of your interesting errands. I am certainly better able to discharge these than requests to buy frippery in foreign countries, which are sometimes wrongly made of me.

I am holding siege, hearing and making the thunder grumble, burning houses and churches, because steeples are good storehouses for bombs; in sum, doing all the evil I can to ones who did me none, and whom I do not know; but that is my trade. I am overwhelmed in my turn with cannonballs, shells, and bombs, and everything goes as well as possible. That obstinate governor of Breslau will perhaps take his leave one day, and it will be good for the city and for us.

I hope that your health is improved and that your parents and your sisters remain well. These are at least the wishes of your devoted servant and admirer.

J. Pernety[18]

Sophie Germain immediately sent off the following letter to Gauss in order to enlighten what must have been a confused genius.

Monsieur,   Paris, 20 Feb., 1807

The consideration due to superior men will explain the care I have taken to ask General Pernety to make it known to whomever he thought appropriate that you have the right to the esteeem of any enlightened government.

In describing the honorable mission I charged him with, M. Pernety informed me that he had made known to you my name. This has led me to confess that I am not as completely unknown to you as you might believe, but that fearing the ridicule attached to a female scientist, I have previously taken the name of M. LeBlanc in communicating to you those notes that, no doubt, do not deserve the indulgence with which you have responded.

The appreciation I owe you for the encouragement you have given me, in showing me that you count me among the lovers of sublime arithmetic whose mysteries you have developed, was my particular motivation for finding out news of you at a time when the troubles of the war caused me to fear for your safety; and I have learned with complete satisfaction that you have remained in your house as undisturbed as circumstance would permit. I hope, however, that these events will not keep you too long from your astronomical and especially your arithmetical researches, because this part of science has a particular attraction for me, and I always admire with new pleasure the linkages between the truths exposed in your book. Unfortunately, the ability to think with force is an attribute reserved for a few privileged minds, and I am sure that I will not encounter any of the developments that you deduce, seemingly so effortlessly, from those that you have already made known.

I include with my letter a note intended to show you that I have maintained an appetite for analysis that the reading of your work has inspired, and that has continually provided me with the confidence to send you my feeble attempts, without any other recommendation to you than the goodwill accorded by scientists to admirers of their work.

I hope that the information that I have today confided to you will not deprive me of the honor you have accorded me under a borrowed name, and that you will devote a few minutes to write me news of yourself. Believe, Monsieur, the interest I attach to this, and be assured of the sincere admiration with which I have the honor to be,

Your very humble servant,
Sophie Germain

[My address is: Mlle. Sophie Germain, chez son père, St.-Crois-de-la-Bretonnerie, number 23, Paris][19]

Gauss responded promptly and at length:

Mademoiselle,   Brunswick, 30 April, 1807

Your letter of February 20, which did not arrive until March 12, was for me the source of as much pleasure as surprise. How pleasant and heartwarming to acquire a friend so flattering and precious. The lively interest that you have taken in me during this war deserves the most sincere appreciation. Your letter to General Pernety would have been most useful to me, if I had needed special protection on the part of the French government.

Happily, the events and consequences of war have not affected me so much up until now, although I am convinced that they will have a large influence on the future course of my life. But how can I describe my astonishment and admiration on seeing my esteemed correspondent M. LeBlanc metamorphosed into this celebrated person, yielding a copy so brilliant it is hard to believe? The taste for the abstract sciences in general and, above all, for the mysteries of numbers, is very rare: this is not surprising, since the charms of this sublime science in all their beauty reveal themselves only to those who have the courage to fathom them. But when a woman, because of her sex, our customs and prejudices, encounters infinitely more obstacles than men in familiarizing herself with their knotty problems, yet overcomes these fetters and penetrates that which is most hidden, she doubtless has the most noble courage, extraordinary talent, and superior genius. Nothing could prove to me in a more flattering and less equivocal way that the attractions of that science, which have added so much joy to me life, are not chimerical, than the favor with which you have honored it.

The scientific notes with which your letters are so richly filled have given me a thousand pleasures. I have studied them with attention, and I admire the ease with which you penetrate all branches of arithmetic, and the wisdom with which you generalize and perfect. I ask you to take it as a proof of my attention if I dare to add a remark to your last letter. It seems to me that the inverse proposition 'If the sum of the nth powers of two numbers is of the form $hh + nff$, then the sum of the numbers themselves will be of the same form' is put a little too strongly. Here is an example of where this rule fails:

$$15^{11} + 8^{11} = 8649755859375 + 8589934592 = 8658345793967 =$$
$$(1595826)^2 + 11(745391)^2$$

Nevertheless $15 + 8 = 23$ cannot be reduced to the form $xx + 11yy$ . . .[20]

Gauss proceeded to elaborate on her faulty proposition, then returned to a description of his own work, presenting her with three new theorems on cubic and biquadratic residues without including the proofs, ". . . in order not to deprive you of the pleasure of finding them yourself, if you find it worthy of your time. . . ." He closed with, "Continue, Mademoiselle, to favor me with your friendship and your correspondence, which are my pride, and be persuaded that I am and will always be with the highest esteem, Your most sincere admirer, Ch. F. Gauss."

Evidently it took Sophie Germain approximately one month to supply the proofs to his three new theorems on cubic and biquadratic residues. She sent these to Gauss, appended to a letter dated 27 June, 1807, along with several new propositions of her own making.

Monsieur,   Paris, 27 June, 1807
I owe you a thousand thanks for the flattering things contained in your last letter. I take them solely as encouragement and, indeed, my greatest ambition will always be not to display my unworthiness of this honor you have bestowed on me in promising to continue a correspondence from which I alone have something to gain.

You have taken the trouble of examining an inverse proposition to the one I had communicated to you and have pointed out the error I have made. I recognize the justice of your observations and thank you very much for troubling to give me this advice. If I did not fear taxing your kindness, I would beseech you to render to me this service in the future. I would always consider it a mark of your benevolence.

How I have enjoyed reading your three theorems on residues! I have searched for demonstrations of them. I add them to my letter in order to have you judge them, since they would appear to me free of error only if they received your approbation.

In the future you could not give me any more greater pleasure than by sending me the basic arithmetic propositions that have succumbed to your thought. In attempting to provide proofs for them, I have developed a way of thinking that for me is full of charm. But the difficulty of this task would be too great if I only had recourse to my intellectual powers. For to tell you the truth, I have wanted to consider the residues of powers greater than quadratic, but I have not been able to penetrate this theory; it remains the subject of my curiosity.

Here, nevertheless, are the few propositions I have arrived at and which I would not dare to send you if I could not count on the indulgence to which you have accustomed me. . . .[21]

Despite the indulgence Gauss had earlier shown, a full six months went by before he chose to respond:

Mademoiselle,   Göttingen, 19 Jan., 1808
In thanking you with all my heart for your last letter and the interesting communications you make in it, I ask a thousand pardons for replying so late. This negligence is for the most part a result of changes in my situation. I have changed my residence in order to accept the

position of professor of astronomy in Göttingen, which had been offered to me a long time ago. I say nothing of the annoying circumstances that have finally influenced me to take this step, nor of the new worries to which I find myself exposed here. I hope that the intervention of the Institute, to which I have had recourse before, will put an end to them. Let us contemplate at present the lovely prospect I have of being able to work, above all on my mathematics, and to publish my findings in the journals of the Society of Göttingen. I have the pleasure of sending you the premises upon which this work is based, which will, I hope, bring you some joy. You will pardon me if at this time I cannot conveniently extend myself to the beautiful proofs of my theorems. I admire the wisdom with which you have been able to grasp them in so short a time. I hope to be able to publish soon the entire theory of which these elegant proofs are a part, along with a host of others. How happy my mathematical affairs make me at a time when I can see nothing around me but unhappiness and despair! Only in the sciences, in the bosom of one's family, and in corresponding with one's dear friends can one find relief and rest from general troubles.

The work on the calculation of the orbit of the planets of which I spoke to you in the last letter is finally in the press. I hope that it will be finished in a few months. I have not spared myself the trouble of translating it into Latin, in order that it will find a larger number of readers.

Remain always happy, my dear friend. The rare qualities of your heart and mind deserve it, and continue from time to time to renew the gentle assurance that I may count myself among your friends, a title of which I will always be proud.

Ch. F. Gauss[22]

This letter is the last from Gauss to Sophie Germain. She was to continue to write to him, but the exchange had come to an end. Although his reasons are not spelled out, Gauss seems to be saying that his professional accomplishments had finally brought him to a position that allowed both concentration on what interested him most (despite some teaching responsibilities) and a promise that he would be able to publish all of his work without difficulty.[23] Under these circumstances he would no longer have time or inclination to oversee the creative attempts of a minor talent who was incapable of keeping up with him.

This is not to suggest that Gauss took Sophie Germain's talents lightly. At a time when few in Europe seemed to have the perseverance, wit or interest to study thoroughly and understand his work in number theory, Sophie Germain had provided support. As he had written to Olbers, "Recently I have had the joy of receiving a letter from a young Parisian mathematician, LeBlanc, who is familiarizing himself enthusiastically with higher arithmetic, and gives proofs that he has penetrated deeply into my D.A.;"[24] Gauss had also kept her letters, sending them to Olbers in May, 1807. In this latter communication, there is evidence that this correspondence stimulated his thinking. "Recently I replied to a letter of hers and shared some Arithmetic [sic] with her, and this led me to undertake an inquiry again; only two days later I made a very pleasant discovery. It is a new, very neat, and short proof of the fundamental theorem of art. 131. . . ."[25]

His regard for her talent is again reflected in another letter to Olbers dated 21 July, 1807. "Lagrange remains quite interested in astronomy and higher arithmetic; he considers the two sample theorems that I shared with you a while ago (in which 2 is a cubic or biquadratic residue of primes) as 'most beautiful and most difficult to prove'. But Sophie Germain has sent me the proofs; unfortunately I have not been able to go through them yet, but I think they are good. At least she has approached the matter from the proper side; they are only a little lengthier than required."[26]

Nevertheless, Gauss's treatment of Sophie Germain left her in a difficult position. She was certainly more dependent than he on the intellectual support of their correspondence. Now that he no longer wished to continue writing, the year 1808 found Sophie Germain in need of, and receptive to, some new focus for her intellect, enthusiasm, and not inconsiderable talent.

Looking back over this four-year correspondence, one can see a personal as well as professional history emerge. At first there was the overly obsequious, but determinedly aspiring, mathematician addressing the acknowledged master, and that master encouraging an inferior. With the discovery of the identity of LeBlanc, Gauss changed his tone, writing with an enthusiasm defined by his awareness of her sex. This is corroborated by a statement in a letter to Olbers: "That LeBlanc is the mere assumed name of a young lady, Sophie Germain, certainly amazes me as much as it does you"[27] and by this comment from Bolyai to Gauss: "You once wrote me of a Sophie in Paris; if I were your wife, I would not be too pleased. Write me more of her."[28]

Of course Sophie Germain was not aware of these letters, but she was familiar with the tone adopted by men unable to consider her professionally, independent of her sex. She replied to Gauss pointedly, thanking him briefly for what had gratified her and adding a lengthy discourse. It is this that Gauss declined to comment on. Nevertheless, in ending their correspondence, Gauss wrote with respect and warmth that mitigated the harshness of the message. Both, however, are belied by the last communication that we have on record from Gauss to Sophie Germain. Delambre, the perpetual secretary of the First Class, was the go-between. How Sophie Germain responded we cannot know.

Mademoiselle,   Paris, Monday, 14 May, 1810

In a letter I have received from Mr. Gauss, I am charged with an errand in which he asks me to engage your opinion. Under any other circumstances, I would not have neglected to eagerly seize this occasion to bring you my homage. But I have just returned from paying my last respects to my wife's mother. This sad event, joined with my other affairs, will not leave me free for some time. For the sum of 500 francs, the value of the medal founded by M. Lalande, which has just been presented to him, Gauss desires a pendulum clock. Here are the terms of his letter:

> In place of accepting the rest, 380 francs, in silver, I would prefer a nice pendulum watch. I do not fix its price: whether it be 60 francs or 300 is the same to me, provided that the watch is elegant enough to be offered as a present to my wife and can serve as a decoration for her room. Perhaps Mlle. Sophie Germain, to whom I ask you to make a thousand compliments, will have the goodness to choose.

It seems to me that it is a clock and not a watch that M. Gauss desires. He has translated the German expression *Pendeluhr* as pendulum watch. It concerns, then, the choice of a clock at a price of between 60 and 300 francs; but he wants it to be elegant and I am afraid that I do not know Mme. Gauss's taste. Please inform me, then, Mademoiselle, if I may hope for your assistance and advice and send me your instructions.

I hope to be more at liberty in a few days to visit you to learn of your cares and your opinion in this matter. Soon thereafter I will work on sending the clock to Göttingen.

Accept the respectful sentiments with which I have the honor to be, Mademoiselle, your very humble and obedient servant.

Delambre[29]

CHAPTER FOUR

SETTING THE PRIZE

The exchange of letters between Gauss and Sophie Germain can stand apart from personal, social, or political circumstances. It made very little difference whether Gauss was responding to a letter from LeBlanc or Germain; what mattered was that he could speak of number theory to another interested mathematician, one who was at least able to understand his thinking. Events of war or of personal life faded into the background; in the foreground are two intellectuals dealing abstractly with matters of profound interest to themselves, but unrelated to events, other people or even the realities of nature.

Gauss had been struggling with the complexities of a full personal and professional life throughout these years. He went on to achieve greatness in the history of science and mathematics – making it difficult now to imagine his insecurity during the time he was in correspondence with Sophie Germain or to recognize the constraints on his life and work. We see these reflected slightly in his letters, but only slightly. Sophie Germain herself was free from the demands of expedience: she was not responsible for earning a living, nor would any professorial post be open to her. She dealt with mathematics purely and abstractly, wherever interest would take her.

Her future was, of course, to be far less glorious than that of Gauss. But in historical terms – in terms of understanding what was happening, and how things happen in a human and social context – her life is probably more interesting. She became involved with the scientific community of the time, but because of her sex always had to remain at an intellectual remove. As a woman, it was difficult for her to meet casually with her colleagues to talk about matters of mutual interest. Every conversation was a formal social event requiring letters of invitation, planning for transportation, requests for permission. Sophie Germain could not stop to chat with friends at meetings of the Institute nor get into serious conversation over cigars and brandy after dinner. In this respect she remained all her life in somewhat the same position *vis à vis* her Parisian colleagues as she had been with Gauss; she was always on the outside, like a foreigner, at a distance from the professional scientific culture. This was no disadvantage for abstract inquiries into number theory: the intellectual community in that area was tiny and untouched by demands that its work have practical relevance. But it definitely affected her later work. For her curiosity lured her from the world of number theory to the world of physics, a leap from pure to applied mathematics that she made with little idea of the chasm that separated them – a chasm both in the demands made on appropriateness in mathematical formulation and manipulation and in the importance of conceptual schemes espoused by established scientists dominating an emerging field.

The new direction that absorbed Sophie Germain's life and interest was provided by M. Chladni. In 1808 he came to Paris and demonstrated the existence of modes of vibration of two-dimensional surfaces. The way in which he did this was ingenious and simple. He would take a glass plate, cover it with powder or sand, hold it with two fingers on opposite sides of the plate, and strike its edge with a bow, which would cause the plate to vibrate and emit a pure tone. The powder would be agitated by the vibrations and would move about until it reached the nodes, where there was no motion. There it would remain. The result was a pattern on the plate of lines or curves designating the mode shape of the vibration caused by the bow.

By varying the place of supports, or the number of supports holding the plate, and by varying the place that the bow struck or the way in which it was struck, one could generate different patterns, corresponding to different pure tones. The number of patterns seemed infinite, the patterns were unique to each circumstance, and the results could be reproduced if one duplicated the conditions exactly. One of the nodal curves always included the support points. How the curves and lines were disposed depended on the shape of the plate, the number and position of support points, and the note that was made by the bow striking differently at the same point. A change in any of these conditions created a new pattern.

Here was an observable natural phenomenon, heretofore unseen, though presumably happening in nature every time a musical note was struck from a flat surface – every time a drum was beaten or a tambourine tapped or a gong rung. And it could happen every time a nail was driven or a shoe was dropped on the floor.

This new insight into things unseen called for an explanation from nineteenth century *savants*, whose great enthusiasm and joy was in explaining phenomena that past ages had left shrouded in mystery. Description would not suffice, i.e., there would be no real satisfaction in saying simply that the powder moved to parts of the plate that were at rest. One wanted an explanation for why those parts remained motionless and especially why certain parts remained at rest under one set of conditions and other parts did so under other conditions.

The question 'Why?' can call for different answers at different times. At the turn of the eighteenth century, it would have been satisfied by a mathematical analysis, i.e., an argument through mathematical symbol and reason that would make comprehensible what M. Chladni had made observable.

This was the task that Sophie Germain undertook. Now, however, she was working both with a physical phenomenon and with a public event, for Chladni's demonstrations were not without notoriety in Paris in that time of relative peace and of Napoleon's great power and ambition. In fact, Chladni's patterns became the occasion of a *prix extraordinaire* offered by Napoleon in 1809. The meaning of this prize and its implications require some explanation, both of the state of prizes for scientific work in Napoleonic France, and of this special prize in particular.

Napoleon had evinced an interest in science and mathematics early in his career. While gaining fame as a military leader, he acquired a seat in the First Class of the Institute through the help of Laplace, a member of the Senate and his friend, and because of his own prestige.[1] No doubt his aspiration to this seat was partly due to the honor it carried rather than from knowledge or ability in matters mathematical. But he cared enough to want the place and to occupy it quite faithfully, even sitting as President of the Institute in 1800, before political affairs absorbed his full attention. His concern for science and mathematics was two-fold: it grew from a real interest and belief in the importance of these fields of human endeavor and from his personal and political ambitions. Napoleon wanted himself and France (for in some respects he seemed to consider them synonymous) to exercise hegemony in all important spheres of life. Politically, that meant military domination. Morally and intellectually, that meant not only aesthetic and philosophical superiority, but also scientific excellence. That he wished to bring all of Europe into the embrace of France intellectually as well as politically is indicated in the statement he made, as President of the First Class, inviting all scientists to compete for prizes offered by the Institute: "Persuaded that everything that might hasten this progress is regarded by enlightened men of every country as a sacred duty, the National Institute hopes that you will give this program all possible publicity, either by inserting it in those journals that appear in your country or by any other means."[2]

Prizes for scientific accomplishment had been offered in France through much of the eighteenth century, and they had been used to encourage work in areas felt by the government to be of national importance. Subsequent to the reorganization of the Academy of Sciences in 1795, the First Class of the Institute was charged with proposing and judging each year one contest in the physical sciences and one in the mathematical sciences. Each prize was established as one kilogram of gold, which in 1803 was set at 3000 francs.[3]

In 1808, the usual procedure for conducting these contests was to elect a commission of four or five from the sixty resident members of the class to choose a problem and establish a program for the prize. Once the commission's recommendation had been accepted by the class, the contest would be announced at its public session in January. Normally two years were allowed for the duration of the contest, with a deadline for entries set as the first day in October. Anonymity of the contestants was assured by the following regulation:

The name of the author should not be put on his manuscript, but only a sentence or saying. If desired, a separated and secret card could be attached in an envelope that would include, in addition to the sentence or saying, the name and address of the aspirant. This envelope would be opened by the Institute only if the entry won the prize.[4]

Once all entries for a given prize had been received, a new commission would be elected to sit in judgment. Their decision would be reached by December, and the results announced, with appropriate ritual and fanfare, again at the January public session. Not all prizes were awarded; frequently no entry was judged worthy even of an honorable mention. In these cases the contest could be prolonged, sometimes with a doubling of the award. No money was awarded for an honorable mention.

These were the conditions governing the usual prizes of the First Class of the Institute. Yet, there were other prizes that Napoleon introduced and supported with government funds. Among them was an annual prize, established by the First Consul in 1802, "for the best experiment made each year on the galvanic fluid."[5] Unlike the prizes in the mathematical and physical sciences where a new program was defined each year, the program for the prize in galvanism was established for all time by this request of Napoleon. The monetary award for this prize was also 3000 francs. In 1808 the prize in galvanism was not awarded.

That same year Chladni came to Paris with two goals in mind. One was to prepare for the publication of a French translation of his 1802 treatise on acoustics by submitting his book to the judgment of the Institute for its review; and the other, not unrelated, to request that a mixed commission of the First and Fourth Classes examine his 'Clavi-cylinder,' a musical instrument of his invention.[6] The First Class acknowledged receipt of his book in October, 1808, and appointed the astronomer Burkhardt to prepare the report. Burkhardt qualified for this commision not, clearly, because of his expertise, but because he was an editor from the First Class for the *Bibliothèque Germanique*, established by the Institute.[7] He was therefore conversant with the state of German science and presumably in a position to review Chladni's work. Evidently, he was not greatly impressed, for the oral report that he made to the First Class in the following month had no impact.[8]

The other commission behaved rather differently. This group consisted of three members from the Class of Mathematical and Physical Sciences and four from the Class of Beaux-Arts. After making a report on the clavi-cylinder, they went on to a critical review of the *Acoustics* and distributed a report that praised Chladni for his ingenious experiments and recommended that those concerned with the advancement of mathematical physics turn their attention toward researching the phenomena of plate vibrations. The First Class received this report, written by Prony, 13 February, 1809.[9]

By this time all funds for contests in the mathematical and physical sciences had been committed for the year 1809. The prize in galvanism, as already mentioned, had not been given in 1808. At a prior session of the First Class,[10] Legendre had suggested a doubling of the prize, but at the Class's meeting of 13 February, 1809, a different disposition was made. A commission on galvanism, consisting of Hauy, Laplace, Halle, Rumford, and Guy-Lussac, made the following proposal to be forwarded to His Imperial Majesty: ". . . that the 3000 francs that were not awarded this year for a discovery in galvanism be employed to encourage the mathematical analysis of the experiments made by M. Chladni on the vibration of resonating plates."[11]

Events then moved rapidly: Napoleon's approval came within a week's time. A commission consisting of Laplace, Legendre, Prony, and Hauy was then elected to draw up a program for the prize; a program that was adopted at the First Class session of 13 March, 1809. At the next weekly meeting, 20 March, it was further decided to ask the Class of French Language and Literature to permit proclamation of the prize program at their next public session, which was to be held in April, a few weeks hence. Otherwise, the announcement of this new competition would be delayed for more than six months.

Finally, on the first Monday in April, 1809, at the public session of the Third Class, this *prix extraordinaire* was announced with the following proclamation:

His Majesty the Emperor and King, who has deigned to call M. Chladni before him and see his experiments, being struck by the impact that the discovery of a rigorous theory explaining all phenomena rendered sensible by these experiments would have on the progress of physics and analysis, desires that the Class make this the subject of a prize that will be proposed to all the learned men of Europe. This new conception of benevolent genius, which animates the grand and profound views of His Majesty for the progress and propagation of Enlightenment, will be received with recognition by all peoples who honor and cultivate the sciences.

The class has thus proposed, for the subject of the prize, the development of a mathematical theory of the vibration of elastic surfaces, and a comparison of this theory with experiments.

The prize will be a medal of gold, valued at 3000 francs. It will be awarded at a public session the first Monday of January, 1812.

Entries will be received only until the first of October, 1811. This term is without exception.[12]

If one looks at the events leading up to this proclamation, one finds more questions than answers. Why did the commission chosen to examine the 'clavi-cylinder' report on the *Acoustics* as well? What suddenly made Chladni's patterns of such profound and immediate interest, when a few months earlier they had not excited such a response? Answers to these questions lead us into speculation, requiring us to piece together the probable, explicated from known facts and future developments.

What seemed to lie behind the public and official promulgation of the *prix extraordinaire* is a personal story of Laplace using his interests and influence to further the career of one of his protégés, Simeon-Dénis Poisson. Laplace had already been influential in the advancement of other scientists of Poisson's generation, notably Jean-Baptiste Biot and Dominique Francois Jean Arago, but helping Poisson in his career proved to be a more complex problem.[14]

Certainly this was not because of any lack of mathematical talent on the part of Poisson. His brilliance had been apparent early when, in 1800, as an eighteen-year-old student at the École Polytechnique, he read a mathematical paper to the First Class of the Institute.[15] During the next ten years, he continued to impress this elite gathering with his work, and became recognized as the peer in mathematics of Laplace and Lagrange.[16] Yet despite his productivity and the honor accorded him, Poisson was not yet a member of the First Class of the Institute as the decade drew to a close. This is probably traceable to the primarily analytical nature of his work. He dealt with the problems and theories of others, and devoted his efforts to perfecting and extending their analytical implications. He had not been concerned with other research that might have qualified him for a seat in the First Class in some section other than mathematics. As each section could seat only six members, and no mathematicians had died during those years, Poisson remained outside the Institute.[18]

Certainly Poisson had done no work that would justify his election to the Physics Section of the First Class, but that was the section he was finally elected to in 1812, following the unexpected death of Malus, one of the academy's younger members.[19] Arago, in his biography of Poisson, summed up the story of Poisson's entry into the Institute as follows:

The public has noted with astonishment the late date of Poisson's entry into the Institute; but this astonishment – is it well founded? Yes! if one recalls that some of his students were admitted before him. Yet the facts are explained very simply without casting suspicion on the spirit of justice from which the Academy has never departed in all cases concerning superior men. The scientific class is divided into sections of six members each. In the nomination process we still adhere to the specializations of these divisions with scrupulous care; thus a geometer would hardly ever get into the Physics Section, an astronomer into the Mechanics Section, etc. Poisson had his place marked in the Geometry Section, and it is solely to the fortuitous order according to which death determines vacancies that we ought to impute the so retarded nomination of our illustrious colleague. Finally, impatient at seeing so eminent a man as Poisson outside their fold, the majority of the Academy ignored the rigidity of the principle and named him to the Physics Section, where he remained until his death.
Laplace, who, from the start, had for Poisson the sentiments of a father, contributed much to this decision, which the subsequent work of our colleague in so many branches of mathematical physics has fully justified.[20]

Arago's citation of Laplace's fatherly concern for Poisson suggests that the *prix extraordinaire* was established with Poisson in mind, as part of Laplace's efforts to be helpful to Poisson and to turn his attention to terrestrial problems of physics that were of interest to Laplace himself. Evidence of Laplace's interest in the elastic behavior of materials at that time is found in a note attached to a memoir appearing in the *Mémoires of the Institute 1809*, as well as in the *Mémoires of the Société D'Arcueil*.

Here Laplace espoused a molecular conceptual scheme as a basis for the analysis of a variety of physical phenomena. He specifically addressed the static and dynamic equilibrium of an elastic lamina:

> In order to determine the equilibrium and movement of an elastic, naturally straight lamina that is bent into an arbitrary curve, it has been assumed that at each point its stiffness is inversely proportional to its radius of curvature. But this law is only secondary and derives from the attractive and repulsive actions of molecules, which are a function of distance.[21]

Laplace was concerned with the one-dimensional beam, the 'linear case' of Euler, with which Sophie Germain had begun her work when she undertook to provide an analysis of the modes of vibration of plates. But he was suggesting a fundamental approach to the problem, a way of modelling an elastic medium as a collection of mutually repelling and attracting corpuscles – a model that would apply equally to the plate problem as well as the beam problem. It is just such an approach that Poisson would employ in his own work on the elastic behavior of plates, the results of which were not made public until his reading before the First Class of the Institute in 1814.

Did Poisson concern himself with the plate vibration problem during the years 1809–1811? It is not difficult to imagine his doing so, guided by Laplace, in the hopes of establishing his competence at explaining a new-found physical phenomenon as well as ensuring the benevolent assistance of this influential senior member of the scientific establishment. Laplace sat on the Commission on Galvanism, which recommended the use of its funds to back the *prix extraordinaire*, and was on the commission that drew up the program for the prize. Laplace, in addition, was on close terms with Napoleon and had himself presented Chladni to the emperor. Yet if Poisson did begin work on the plate problem in 1809, he was not able to complete an analysis worthy of the award and he did not submit a memoir in competition.

In many ways Poisson provides an appropriate parallel and contrast to Sophie Germain, both in 1809 when their work seemed to cross for the first time, and later, when Poisson was to enter the area of elasticity, defending analysis within a molecular scheme of understanding, while Sophie Germain was to deny the necessary validity of this conceptual framework and argue for her own viewpoint.

In 1809, both Sophie Germain and Poisson were at the beginning of their careers. Both owed their choice of profession to some extent to the political upheavals of revolutionary times. Sophie Germain's interest in and pursuit of mathematics were tied to a deliberate withdrawal from her predictable social role; the flux of revolutionary society made it possible for her to gain at least some advanced mathematical training. Poisson was no less dependent on politics. The reordering of French higher education undertaken during the revolutionary years was designed to enable men of promise who lacked social or financial status to rise in the learned professions. Poisson was both from an impoverished family and clearly brilliant. With the educational system encouraging just such a youth, he could aspire to a career in mathematics with more realistic expectations than would have been possible at an earlier time.

Yet a career in mathematics had a different meaning for Sophie Germain than it did for Poisson. For Sophie Germain it represented a way out of the contemporary world, a way to partake in an historical tradition. The prize was not a clear step toward anything, neither means toward professional recognition nor needed financial aid (there is no evidence that Sophie Germain lacked money nor indeed that she had costly tastes, save for the latest mathematical works).

For Poisson, on the other hand, a career in mathematics was intimately connected to both honors and money. His first appointment at the Ecole Polytechnique upon graduation in 1800 as *Répétiteur*, and in 1802 as a provisional professor of mathematics, ensured him of an income of a few thousand francs, which was adequate but not generous. The triumphant reception of his memoir on the stability of the solar system brought him an appointment as an adjunct member of the Bureau of Longitudes, an appointment for which Laplace, the director, was responsible. The next year, 1809, saw his publication of a memoir on the rotation of the earth, and his appointment as Professor of Rational Mechanics at the newly created Faculty of Science (4000 francs per year). On the strength of these analytical memoirs concerned with heavenly motions, he was awarded the Lalande Medal for the year 1809 (571 francs).[22]

The patronage of established *savants* also meant something very different for Sophie Germain than it did for Poisson. She had depended on them for her education and for encouragement in continuing her work. Yet she had no professional or social stake in their support. Somehow, her dependency was quite purely intellectual, lacking even a very personal component, since it is clear that she could as easily work as LeBlanc as under her own name. When she became interested in the plate problem, however, patronage took on a different character, as will become clear from the somewhat ambivalent patronage of Legendre

when he, along with Laplace, Lagrange, Lacroix, and Malus, was elected
on 21 October, 1811, to judge the sole entry to the contest for the *prix
extraordinaire*.

CHAPTER FIVE

THE ONE ENTRY

Up until this time Mademoiselle Germain had not published a thing. Then there occurred a remarkable event that established her as an author. A German scholar, Chladni, came to Paris to repeat his intriguing experiments on the vibration of elastic plates. They caused a sensation. Napoleon, before whom they were performed, took a lively interest in them, regretted that they had not been subject to calculation, and, in order to facilitate this, proposed that the Institute offer a *prix extraordinaire*. Yet geometers were completely discouraged by a word from Lagrange, who said that the solution of this problem would require a new kind of analysis. Despite the imposing authority of this geometer of Turin, Mademoiselle Germain never despaired of success. She studied these phenomena in a thousand ways, applied analysis to them, and submitted a memoir in competition wherein she gave an equation for elastic surfaces.[1]

Before Sophie Germain could begin to "study these phenomena in a thousand ways" she had to find a solid framework from which to begin. Such a base is generally taken for granted by members of a profession; it consists of knowledge and methodology so well integrated into one's professional understanding that it can be relied upon without special attention. Further, this basic knowledge is common to all members of the profession, so that the premises from which one member works are shared by all: one needs neither to spell them out nor, indeed, even bring them into consciousness.

Sophie Germain had no such well-formed working knowledge: her education had been too haphazard and her continuing study too limited and random. Her vision of a mathematical methodology appropriate to the explanation of physical phenomena was shaped first from her early encounter with number theory, then from Lagrange's *Mécanique Analytique*,[2] and finally from laboriously translated scraps of Euler's research on the vibration of elastic beams,[3] encouraged and aided by some correspondence with Legendre. However much this kind of educational experience hampered her mathematical performance, it did leave her free to choose, conjecture, and manipulate far more freely than those intimately attached to contemporary scientific reality.

Lagrange's work in mechanics had first appeared in 1788 (a second edition was published in 1811). His *Mécanique Analytique* established a general method for the analysis of a variety of problems, including static and dynamic behavior of fluids as well as solids. Before this time

40

mechanics had relied on diverse and disconnected methodologies, which Lagrange then synthesized into, as he claimed, "some general formulas whose simple development gives all the equations necessary for the solution of each problem. . . . This work . . . unites and presents the different principles previously found to facilitate the solution of problems in mechanics, showing their relationship and mutual dependence, and permits one to judge their appropriateness and applicability."[4]

It seems improbable that Lagrange, in saying (if indeed he did) that the plate problem called for a "new kind of analysis," meant a new *Mécanique Analytique*. More likely, he had recognized that, given an appropriate model from which to derive a differential equation, any solution of this equation would depend on one's considering two spatial dimensions, which would lead the would-be solver into regions uncircumscribed by contemporary analytical technique. Problems arising from two spatial dimensions did not, however, deter Sophie Germain. Her knowledge of the calculus was not really sophisticated enough for her to be frightened away from any maze of partial differentials and multiple integrals.

Thus she started work on the problem of the vibration of plates, beginning her research reasonably enough with Euler's investigation of the vibration of beams, the one-dimensional, or, as she called it, the linear case. She was directed to this work by the statement announcing the prize itself, and could expect to learn from Euler principles and methods applicable to the plate problem. Later she wrote,

As soon as I learned about M. Chladni's first experiments it seemed to me that analysis could determine the laws that govern them. But I chanced to learn from a great geometer [Lagrange] whose first works had been devoted to the theory of sound that this problem contained difficulties that I had not even suspected. I stopped thinking about it.

Seeing M. Chladni's experiments during his stay in Paris excited my interest anew. I began studying Euler's memoir on the linear case, certainly not with the intention of competing for the *prix extraordinaire* proposed by the Institute, but only desiring to come to appreciate those difficulties that the terms of the program brought to mind.[5]

A flurry of correspondence with Legendre indicates that she undertook the study of Euler with vigor, even though her approach to the prize was more of an inquisitive than competitive nature at that time. We do not have to rely simply on her word for this judgment: by January 1811, a full year and a half after the announcement of the competition, when her correspondence with Legendre on this subject ceased, she had done nothing resembling an analysis of plate vibration modes. Within the following eight months she did all the work that led to her entry in the competition.

Obviously, her way of undertaking research was entirely different from what we would expect of a professional in the field – of a Poisson, for instance. At the beginning it was that of a dilettante, a dabbler who was trying to understand and appreciate what others were doing. This was the role naturally assigned her by society. Yet her seriousness and natural talents went beyond those of the dilettante, so that finally she emerged into the professional world.

First, however, she investigated Euler. If we are to follow her later work, it seems wise to begin where she did. In his memoir, Euler first deduced from mechanical principles an integro-differential equation that expressed the moment equilibrium of the portion EY of the lamina or beam EF shown in Figure V.1:[6]

$$\int dy \int P ds - \int dx \int Q ds = V/r.$$

The left-hand side of this equation represents the effect of external forces acting on the beam: P acting in the horizontal direction, Q in the

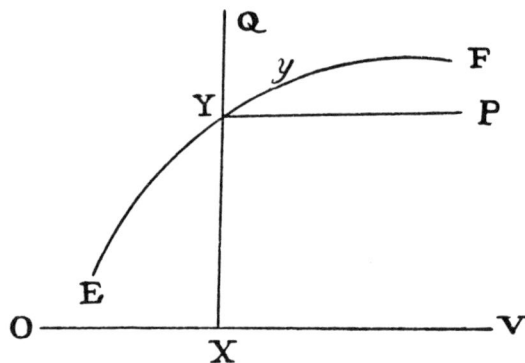

Fig. V.1   Euler's elastic lamina.

vertical. They tend to bend and displace the lamina, producing a radius of curvature r. V, the "absolute elasticity" is a constant which depends on the beam's material substance. In effect, the right-hand side represents the internal forces which balance, or counteract, those applied externally.

Euler considered the case of a horizontal beam experiencing small displacements in the vertical direction, with negligible horizontal motion. A succession of differentiations of the above equation yielded the following differential equation:

$$\frac{1}{2g}\left(\frac{ddy}{dt^2}\right) = -bcc\left(\frac{d^4y}{ds^4}\right).$$

Euler's differential equation defines the behavior of each element, or point, along the beam. Neither he nor anyone else had derived a corresponding equation for a plate – a similar case, but with two spatial dimensions. It was this lack that the statement of the prize acknowledged in saying, "Thus we do not even possess the differential equations of movement (of the plate) for this type of vibration in considering these phenomena as nature presents them."

From his equation for the behavior of each element in a beam Euler went on to derive solutions to describe how the beam moves as a whole. He sought a solution that would enable him to compute the displacement of the beam at any position s and at any time t.

The differential equation as it stands admits of solutions in a variety of different forms. Euler considered a special class of solutions, namely those defining a "regular vibratory motion," solutions wherein each point along the beam moves up and down in simple harmonic motion, like the to-and-fro motion of the pendulum.

Such solutions are described by the following functional relationship:

$$y(s,t) = \sin\left(\zeta + t\sqrt{(2g/k)}\right)\{\alpha e^{s/f} + \beta e^{-s/f} + \gamma\sin(s/f) + \delta\cos(s/f)\}$$

where f is related to the frequency of vibration $\sqrt{(2g/k)}$, by

$$1/f^2 = \text{constant} \cdot \sqrt{(2g/k)}.$$

To determine the possible values of f, that is, the resonant frequencies of vibration, and to establish the four parameters, α, β, γ, δ that fix the associated mode shapes, it is necessary to specify conditions at E and F, the ends of the beam. (See Figure V.2.)

Euler considered various cases of end conditions. Two were of interest to Sophie Germain:

(1) (listed in his memoir as Case IV) wherein the beam is fixed at its extremities, E and F, by a "stylus." Here the points E and F are fixed in space, but the beam is free to rotate about one axis at these points, as if these ends were hinged to a support. He demonstrated that the spectrum of discrete resonant frequencies is:

$$\frac{\pi c\sqrt{2gb}}{a^2}, \frac{4\pi c\sqrt{2gb}}{a^2}, \frac{9\pi c\sqrt{2gb}}{a^2}, \ldots, \text{etc.}$$

That is, it is proportional to the sequence $1, 2^2, 3^2, \ldots,$ etc.

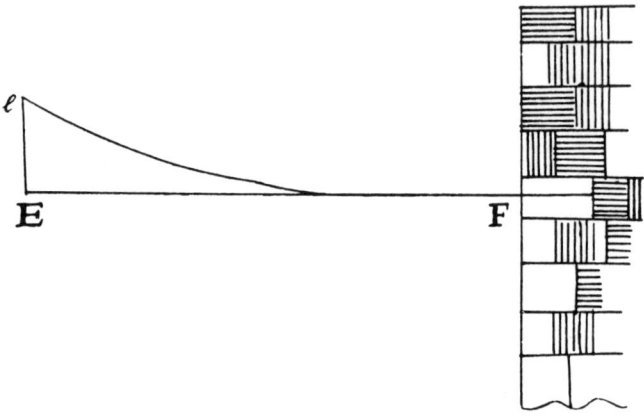

Fig. V.2   A free-fixed elastic lamina.

(2) (Listed in his memoir as Case V) where the beam is simply fixed at E but rigidly fixed at F, meaning that at point F the beam is restrained from rotation as well as displacement, as if this end were mitred into a wall. In this case Euler showed that the spectrum of discrete resonances is, approximately:

$$25\frac{\pi c\sqrt{2gb}}{16a^2}, 81\frac{\pi c\sqrt{2gb}}{16a^2}, 169\frac{\pi c\sqrt{2gb}}{16a^2}, \ldots, \text{etc.}$$

that is, it is proportional to the sequence $5^2, 9^2, 13^2, \ldots,$ etc.

In both cases he then deduced the relative values of the parameters α, β, γ, δ defining the mode shape associated with each of these frequencies.

Euler then proceeded to a problem that drew Sophie Germain into the complexities of this kind of analysis. It is a variation on the previous problem. The beam was assumed to be restrained by hinges at both its ends, and again by a hinge at a third point arbitrarily located along the span. (See Figure V.3.) Thus when total length of the beam is defined as a, a stylus capable of preventing displacement but not rotation was considered to be located at point L, a distance λa along the beam. Again the problem was to determine the frequencies of vibration and associated mode shapes.

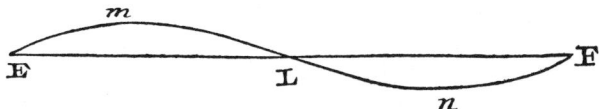

Fig. V.3   An elastic lamina with stylus support at L.

Euler demonstrated that all possible frequencies of vibration could be determined from the roots of the equation:

$$0 = 2 - 2e^{2\omega} - (e^{\lambda\omega} - e^{\omega(2-\lambda)})\,(e^{\lambda\omega} - e^{-\lambda\omega})\,\frac{\sin\omega}{\sin\lambda\omega\cdot\sin(1-\lambda)\omega}$$

where, strictly for convenience's sake, ω has been introduced in place of a/f. Determination of those values of ω that make the right-hand side of this equation vanish will yield values of f and, hence, the frequencies of vibration.

Euler worked through the solution of this problem only for the special case wherein the stylus was located at midspan, i.e., $\lambda = 1/2$. Here, the equation, from whose roots the frequencies of vibration can be determined, becomes:

$$0 = 2 - 2e^{2\omega} - (e^{\omega/2} - e^{3\omega/2})\,(e^{\omega/2} - e^{-\omega/2})\,\frac{\sin\omega}{[\sin(\omega/2)]^2}$$

which, after some manipulation may be written:

$$0 = [\sin(\omega/2)]\,[(1 + e^{\omega})\sin(\omega/2) - (e^{\omega} - 1)\cos(\omega/2)].$$

He then pointed out that either factor in this expression, on being set equal to zero, yields a solution. He considered these two solutions separately. In what follows the spectrum of frequencies determined from setting the first factor, $\sin(\omega/2)$ equal to zero will be called solutions of the 'first kind;' that spectrum determined from setting the second factor $(1 + e^{\omega})\sin(\omega/2) - (e^{\omega} - 1)\cos(\omega/2)$ to zero, will be called solutions of the 'second kind.'

Euler wrote:

For the first case, if we set $\sin(\omega/2) = 0$ then $\omega/2 = i\pi$ where i denotes an arbitrary integer. The result is that innumerable regular motions exist. It is evident that this case agrees entirely with the fourth case derived above apart from the fact that here $\omega/2$ was there ω. Of course, here the length of either portion is $a/2$. Either half, therefore, executes its vibrations in the same manner and as though it existed independently and was fixed by a stylus at its

ends; for which reason, all simple tones which either portion can emit would be expressed by
the following numbers:

$$4 \frac{\pi c \sqrt{2gb}}{a^2}, \ 16 \frac{\pi c \sqrt{2gb}}{a^2}, \ 36 \frac{\pi c \sqrt{2gb}}{a^2}, \ \ldots$$

which are therefore higher by two octaves than in Case IV. The reason for the difference is
that here the length is now half that of the other case . . .[7]
    In addition, the second solution arising from the other factor yields:

$$\tan(\omega/2) = \frac{e^\omega - 1}{1 + e^\omega} = \frac{e^{\omega/2} - e^{-\omega/2}}{e^{\omega/2} + e^{-\omega/2}}$$

which is the same as the result obtained in Case V. The only difference consists in this: here
$\omega/2$ appears, whereas there ω appeared. We learn, therefore, that either portion EL or LF
can shake violently in the same way as if each portion is simply fixed at E or F by means of a
stylus while firmly fixed at L. The values of ω in this case will be $5\pi/2$, $9\pi/2$, $13\pi/2$, $17\pi/2$,
$21\pi/2$, etc., and sounds will originate two octaves higher than in the fifth case.
    It is thus worthy of our greatest attention that the two portions EL and LF can vibrate in a
double mode: in one manner like the fourth case, considered above, and in the other
manner, in a way that agrees with the fifth case. In addition, the values of the coefficients can
be obtained in the same way as they were derived above, except that, in this case, we cannot
set $\sin(\omega/2) = 0$.[8]

Sophie Germain attempted a generalization of Euler's analysis by
admitting other possible positions for the internal stylus. Euler
had carried his analysis through to obtain numerical results only for
the special case where the stylus was positioned at midspan, i.e.,
$\lambda = 1/2$. Sophie Germain tried to find solutions for any rational
λ.[9]

    Following Euler's procedure for the solution of the 'first kind,' she
re-wrote the equation Euler had derived as:

$$0 = \sin\lambda\omega \ \sin(1 - \lambda)\omega \ (2 - 2e^{2\omega}) - (e^{\lambda\omega} - e^{\omega(2-\lambda)})(e^{\lambda\omega} - e^{-\lambda\omega})\sin\omega.$$

A discrete spectrum of values of ω that makes both $\sin\lambda\omega$ and $\sin\omega$ vanish
then satisfies this relationship. For example, if $\lambda = 1/3$, that is if the stylus
is positioned a third of the way down from the span from point E, each
one of the spectrum of values

$$\omega = 3\pi, \ 6\pi, \ 9\pi, \ \ldots, \text{ etc.,}$$

satisfies the equation, and hence yields a possible frequency of vibration.

    But then, when she sought solutions of the 'second kind' for rational
values of λ, assuming $\sin\lambda\omega$ not equal to zero, she encountered difficulties
that were not apparent from Euler's derivation. Therefore she
approached Legendre in January of 1811, presenting him with her

analysis, questioning him about this difficulty, and requesting a critique of her proposed solution of the 'first kind.'

Legendre responded with the following note. Here he took Euler as well as Sophie Germain to task for their solutions of the 'first kind,' then proceeded to describe a general procedure for determining the spectrum of values of ω for solutions of the 'second kind.'

(page 155)[10]

The equation $\sin(\omega/2) = 0$ does not necessarily follow from the equation to be solved; it comes from the introduction of a factor through multiplication and is extraneous to the solution of this problem.

In fact, the first form of the general equation (page 154) is:

$$0 = 2 - 2e^{2\omega} - (e^{\lambda\omega} - e^{\omega(2-\lambda)}) (e^{\lambda\omega} - e^{-\lambda\omega}) (\cot\lambda\omega + \cot(1 - \lambda)\omega).$$

If one puts $\lambda = 1/2$, it becomes:

$$0 = 2 - 2e^{2\omega} - (e^{\omega/2} - e^{3\omega/2}) (e^{\omega/2} - e^{-\omega/2}) (2\cot(\omega/2)).$$

Now, the supposition '$\sin(\omega/2) = 0$' does not satisfy this. It will only be satisfied when $\omega/2 = 0$, or $\omega/2$ is infinitely small, which is not a real case, but an abstraction.

Besides, the solution $\sin(\omega/2) = 0$ is inadmissable because it yields infinite values for the coefficients on page 153 (always taking $\lambda = 1/2$). Euler ought to have mentioned this when he stated (page 156), "We multiply all coefficients by $\sin(\omega/2)$." One can easily multiply the ordinate of a curve by a constant in order to render the curve sensible through a geometrical construction. But one cannot multiply by zero.

Thus it is the solution of the 'second kind' that is legitimate and that I find unobjectionable.

When Mlle. Sophie wished to consider the general case, it seems to me that she fell into the same error as Euler in taking $\sin\lambda\omega = 0$. This solution is an illusion; it results from improperly introducing a factor into the equation. Further, as in the case $\lambda = 1/2$, it would result in the disadvantage of making the coefficients of the curve $\alpha, \beta, \gamma, \delta$, etc., infinite.

As for the rest, [the solution of the 'second kind'] except for the smallest value of ω, which would require some trial and error before one could fix its precise value, in general it is easy to solve Euler's equation on page 154, i.e.,

$$0 = 2 - 2e^{2\omega} - (e^{\lambda\omega} - e^{\omega(2-\lambda)}) (e^{\lambda\omega} - e^{-\lambda\omega}) (\cot\lambda\omega + \cot(1 - \lambda)\omega)$$

In fact, if one understands well the spirit of the solution for the six principal cases, one sees that except for the smallest value, and sometimes even then, the quantity e^{ω} becomes so large that one can neglect $e^{-\omega}$ with respect to e^{ω} with complete confidence, just as one can neglect $e^{-\lambda\omega}$ with respect to $e^{\lambda\omega}$. According to this principle, the preceding equation reduces to:

$$0 = -2e^{2\omega} + e^{2\omega}(\cot\lambda\omega + \cot(1 - \lambda)\omega)$$

or simply:

$$\cot\lambda\omega + \cot(1 - \lambda)\omega = +2$$

Now, setting $\cot\lambda\omega = x$; $\cot(1 - \lambda)\omega = y$ one easily finds, for different values of ω, an algebraic equation relating x and y, which together with the equation $x + y = 2$ will give a finite number of solutions. For example:

$$\lambda\omega = \alpha, \lambda\omega = \beta, \lambda\omega = \gamma.$$

From these solutions, one then forms the general solution:

$$\lambda\omega = \alpha + \kappa\pi, \lambda\omega = \beta + \kappa\pi, \lambda\omega = \gamma + \kappa\pi, \text{etc.}$$

κ being any number you wish.

Thus, there will exist as many kinds of values for ω as roots of the equation in x. For example, letting $\lambda = 1/3$, it is necessary to satisfy the equation

$$2 = \cot(\omega/3) + \cot(2\omega/3).$$

Now if one sets $\cot(\omega/3) = x$ one will have $\cot(2\omega/3) = (x^2 - 1)/2x$ from which $x + (x^2 - 1)/2x = 2$ or $3x^2 - 1 = 4x$ thus $x = (2 \pm \sqrt{7})/3$. Let us call α and β the two angles between 0 and 180° which yield $\cot\alpha = (2 + \sqrt{7})/3$ and $\cot\beta = (2 - \sqrt{7})/3$ and we will have in general:

$$(1/3)\omega = \alpha + \kappa\pi, (1/3)\omega = \beta + \kappa\pi.$$

That is, the values of ω will form two distinct series:

$3\alpha, 3\alpha + 3\pi, 3\alpha + 6\pi$, etc.
$3\beta, 3\beta + 3\pi, 3\beta + 6\pi$, etc.

each holding for a mode in which the beam may oscillate.

In practice, it would be necessary to search more carefully for the exact values of the first two terms $3\alpha, 3\beta$; but the others will always be sufficiently close."[11]

Sophie Germain could not accept Legendre's critique of her solution of the 'first kind,' nor should she have. Despite the awkward process employed by Euler in setting $\sin(\omega/2) = 0$ and in obtaining the relative values of the parameters $\alpha, \beta, \ldots$, etc., his solution, and hers as well, is correct. In defense of her demonstration she evidently argued that her solutions could be considered a special subset of solutions to Case IV, i.e., those modes of vibration of a beam, simply fixed at its ends, that have a node, a point of zero displacement, at the point where the stylus is positioned. These solutions appeared, clearly, to be appropriate. On the other hand, she encountered difficulties when she sought solutions of the 'second kind.' If the stylus was placed somewhere other than at the midpoint of the span, then Euler's interpretation of his results for $\lambda = 1/2$ in terms of solutions to Case V would no longer be appropriate. For arbitrary λ, one could not, in general, have the beam vibrating as if it were two beams, each of a different length, and each behaving as if it were rigidly fixed at the stylus and simply fixed at its other end because the spectrum of Case V does not permit a consonance of frequencies for beams of different length under these end conditions.[12]

In responding to Sophie Germain's reply, Legendre admitted his error in objecting to solutions of the 'first kind.' He then proceeded to reveal a real error in Euler's work which, evidently, Sophie Germain too had discovered. In closing he hastily dismissed her objections to Euler's solutions of the 'second kind.'

Paris, 19 January, 1811

The multiplication by $\sin(\omega/2)$ that I objected to in my first note becomes clear when one examines things more closely. Here is my comment:

Before making any supposition about the value of ω, the author (page 154) deduced the relationship

$$\frac{\alpha}{\alpha'} = \frac{e^{\lambda\omega} - e^{(2-\lambda)}}{e^{\lambda\omega} - e^{-\lambda\omega}}$$

from which he concluded:

$$\alpha = e^{\lambda\omega} - e^{\omega(2-\lambda)}, \alpha' = e^{\lambda\omega} - e^{-\lambda\omega},$$

since, in fact, he is free to multiply all coefficients α, β, γ, δ by the same number; there always remains an arbitrary coefficient C multiplying all of them. But when one sets $\omega/2 = 0$ one finds $\alpha = 0$, $\alpha' = 0$. It follows that one has improperly multiplied all coefficients $\alpha, \beta, \gamma, \delta$ by an indeterminate quantity, since a coefficient α, which without this multiplication would have been zero, becomes a finite quantity.

To rectify this error then requires suppressing the infinite factor, or multiplying by $\sin(\omega/2)$.

This explanation is still somewhat obscure and it is much simpler to redo the calculation of the coefficients in supposing that $\sin(\omega/2) = 0$ or $\omega/2 = \nu\pi$. λ being 1/2.

Thus, lettting $\sin(\omega/2) = 0$ and going back to the first equations on page 152, one easily finds that $\alpha = 0$, $\beta = 0$, $\delta = 0$, $\alpha' = 0$, $\beta' = 0$, $\delta' = 0$. Only γ and γ' do not vanish. But these equations do not determine their values uniquely; one finds simply $\gamma = \gamma'$. Since one can introduce a common factor C, one can then set $\gamma = 1$ and one will have $\gamma' = 1$. Euler, in his analysis (unusual for him) found $\gamma' = -\gamma$, but there is clearly an error; equations III and IV on page 152 evidently give $\gamma = \gamma'$.

Here is a very grave and real difficulty. Witness the consequences that result from this. If one has $\gamma = \gamma'$ then the equation of the portion LF of the curve (page 157) is no longer

$$y = -C\sin(u\omega)$$

but rather

$$y = +C\sin(u\omega)$$

the same as portion EL.

It remains then to determine which of the equations is true. At first glance, it might appear that Euler implies that all of the ordinates are negative for the portion LF. This is not the case. Euler has been misled by the equation because of his error in the sign of γ' and the true equation of portion LF is:

$$y = C(\zeta + t\omega^2 c\sqrt{3\beta/a^2}) \sin(u\omega)$$

which is absolutely the same as that for portion EL, that is, the two portions are described by the one and the same curve designated by the same equation: this result completely reconciles the theory that Mlle. Sophie wishes to adopt, despite Euler's equations and despite my first note.

All one needs to be convinced of this is to observe that since $\sin(u\omega)$ is zero when one sets $u = 1/2$, the two suppositions $u > 1/2$ and $u < 1/2$ will give results of opposite sign for y, so that before and after the point L, the ordinates will be of different sign.

Thus the difficulty on this point is entirely resolved; it comes from the error in sign that Euler made in the equation $\gamma = -\gamma$!

I ought to add that in contradistinction to the opinion I advanced in my first note, the factor $\sin(\omega/2)$ gives the admissable solution of $\sin(\omega/2) = 0$ or $\omega/2 = \alpha\pi$.

With respect to the other solution [the solution of the second kind] against which I saw no objection, it appears to me that one cannot reject it simply because the sounds emitted by the beam in the two portions are not in harmony with each other. Oscillations can very well take place without being harmonic.

Believing, Mademoiselle, that you should not be made to wait until Monday for these explanations, I send them to you as a proof of my zeal and devotion.

Legendre[13]

Justified in her approach which yielded solutions of the 'first kind,' Sophie Germain apparently still held the solutions of the 'second kind' suspect, and wrote again to Legendre questioning whether they were perhaps only analytic, i.e., quirks of mathematical process that had no basis in physical reality.

Legendre's response showed that he had now read Euler's memoir with care. He went directly to the root of the problem with solutions of the 'second kind,' and left Sophie Germain with both a valuable explication of why such solutions can differ from particular solutions for the beam supported only at its ends (Case IV) and a clear indication that he had had enough of the whole problem.

28 January, 1811

Euler has treated his problem of §47 casually and limited himself to one example. His mistakes could be rooted as much in his calculations as in his reasoning. He is certainly mistaken in his calculations when he deduces $\gamma = -\gamma'$, since one ought to have $\gamma = \gamma'$. It could also be that the second solution is purely analytic and does not satisfy the physical circumstances of the problem. I have not made up my mind on this, not having sufficiently reflected on these types of questions and not having the leisure or the taste to devote myself to a more profound examination. I prefer to give up the cause to Mlle. Sophie and to admit defeat, rather than struggle with her on a subject that she has thought so much upon.

Before beginning our discussion, it is necessary to define precisely the word *simpliciter fixus* which Euler uses. Since at such a point y is always zero, it seems to me necessary to regard the "stylet" as a fixed needle that traverses the rod at the middle of its length and about which it can turn in all directions. I cannot see that this word of Euler's can have any other meaning.

Having granted this, and having already having determined in problem IV all the motions that an elastic rod whose extremities are simply fixed can take among all these regular, possible motions, there would be a certain number in which the middle point of the rod would be at rest. These latter movements would be solutions to the problem of §47; it is only a question of weeding out from the general solution of problem IV all the solutions that do not satisfy this condition.

Similar reasoning applies to all other general cases from problem I to problem VI and even applies when the "stylet" is positioned at some other point than at the middle, or even to the case where several supports are positioned at different points along the beam, at least if the distances are a rational ratio of the complete length of the beam.

This explanation makes many difficulties disappear, but I cannot hide the fact that it is open to objection.

When one considers in the succession of problems I, II, . . . VI, the different motions of the beam, one supposes that it is completely free at these intermediate points, and that they experience no resistance in the course of their motion. The situation is no longer the same when one introduces one or several supports applied at different points. If these supports do not carry any load in any direction, the solution as we have come to conceive it may be applied; but if they support a load, one must take this into account. The solutions to problems I, II, . . . VI are no longer applicable and our entire edifice crumbles.

Permit me, mademoiselle, to leave to you the task of freeing yourself as you can from these ruins. As for me, I excuse myself, in offering you my most humble reverence.

Legendre[14]

This Eulerian exercise and correspondence with Legendre served to engross Sophie Germain in the problem of vibration and to provide her with a framework for inquiry. Within the next eight months, left to her own devices, she fashioned a memoir and submitted it in competition to the First Class on the twenty-first of September. In the next month Legendre, Laplace, Lagrange, Lacroix, and Malus were elected to judge the competition. There was only this one entry.

On the frontispiece of her treatise she affixed the following identifying quotation:

"Effectuum naturalium ejusdem generis eaedem sunt causae"
(Newton, *Phil. Nat. Prin. Math. . .Philos. Regula* II)[15]

Hidden away in an attached envelope was the quotation again and the name of Sophie Germain. From the letter that follows it is evident that at least one of those elected to judge this contest knew the name contained in that envelope. Probably Sophie Germain had written Legendre shortly after the deadline for entries informing him of her effort and asking whether it had been received. Again, only Legendre's response appears to have been preserved.

Mademoiselle,   Paris, 22 October, 1811
    Your memoir is not lost; it is the only one that we have received concerning the problem of the vibration of surfaces. Yesterday we designated five commissioners to examine it. I have the honor of being one of them. M. Laplace, Lagrange, Lacroix, and Malus are the other four. I have said nothing; I advise you, as well, to keep silent until a definite judgment is made.
    I am, with all the sentiments that you know, your devoted servant.
                                                                      Legendre[16]

This short note reveals a number of things – first and most obviously, that the rules designed to preserve secrecy were not being followed with any great seriousness. Although Legendre enjoined Sophie Germain to silence, he did not feel it necessary to disqualify himself from judgment because of his illegal knowledge. Further, this note testifies to a level of friendship and trust between them that is not clear to outsiders from their previous correspondence. Sophie Germain could not have approached Legendre in a purely intellectual way about her memoir; she had to trust him as a friend. His reply accepts this trust and reveals him, to some degree, as a co-conspirator.

    Their violation of the rules of the competition went even further, however. (Perhaps it merits mention that in most cases we would have no evidence of such rule-breaking, since it would occur orally between friends who saw each other easily.) The following month, in November, we find the following letter from Legendre:

Mademoiselle,   Paris, 10 Nov., 1811
    Your memoir is being circulated. M. Lacroix had it in his hands last Monday. I will find out tomorrow whom he has passed it on to, and I will add to it the supplement. The commissioners will then judge whether to take account of this supplement or not. Moreover I will see to it that M. Lagrange does not delay in reading the whole thing.
    There is no difficulty, it seems to me, in the particular case where the pendulum has the velocity required to climb to the extremity of its vertical diameter. The calculus proves that it requires an infinite time for the pendulum to arrive at this point, and there its motion will vanish.
    Agree, Mademoiselle, to the respect of my most distinguished sentiments.
                                                                      Legendre[17]

Although it is perfectly clear what Legendre is responding to, it is difficult to determine exactly what Sophie Germain was doing from October to

November. Did she submit an essentially unfinished memoir in October in order to meet the deadline, counting on Legendre's indulgence to be able to add a supplement to it? This seems possible since he was not surprised at receiving it. Or, having completed her work, did she begin to suffer doubts while waiting, moving her to fashion a supplement? When one considers the daily reality of her life, the latter seems more likely. So few demands were made on her time or energy that, having completed a difficult and engrossing piece of work, she was left with a feeling of great emptiness having to wait an indefinite period of time until a decision on her work was made. The following *pensée* she once jotted down reflects something of her feelings and of the spirit in which she tried to understand existence. It also deals with non-intellectual, purely personal responses such as impatience:

Space and time: these man proposes to measure. The one circumscribes his momentary existence, the other accompanies his successive stages in life. These two dimensions are tied together through a necessary relationship, namely, motion. When motion is constant and uniform, space is known by time and time is measured by space. Man has nothing within him that is constant and uniform; continually modified every instant, he is changing, irregular, and hardly durable enough to be a measure of duration. [18]

She had spent a month continuing her work on her memoir, knowing that her hastily done analysis was wanting in at least some particulars. Thinking that she had developed a convincing argument for one of the more questionable parts of her work, she sent Legendre a supplement with the trust that he would do what he could for her. But she was not completely comfortable writing him with a bare-faced request for collusion. She included in her letter a question about a pendulum problem, which she would have found again in Euler, whose work she was evidently still consulting rather closely. That Legendre answered her question so simply, even off-handedly, testifies to its superficiality. Evidently it was something she had come upon and was curious about, and directing an inquiry to Legendre about it served to soften the impropriety of her letter to him. He, however, seemed quite unembarrassed; quite likely such favors were common among colleagues and friends.

Two months remained before January's public session would bring an announcement of the outcome of the competition. Legendre, however, wrote her in December, allaying the anxiety aroused by waiting, though at the same time dampening out whatever hopes she had of success:

Mademoiselle,   4 Dec., 1811

    I do not have good news to give you concerning the examination of your memoir. Your principal equation is not correct, even assuming the hypothesis that the elasticity at each point can be represented by $(1/r) + (1/r')$. M. de Lagrange has found that, using this hypothesis, the correct equation should be of the form:

$$\frac{d^2z}{dt^2} = k^2 \left(\frac{d^4z}{dx^4} + 2\frac{d^4z}{dx^2dy^2} + \frac{d^4z}{dy^4} \right)$$

assuming also that z is very small. I have hardly verified this calculation; one can consult the author. But what confirms its correctness at first sight is this: supposing that the vibrating surface is reduced to a beam of constant size, which can be expressed by making $dz/dy = 0$, one returns to the equation $(d^2z/dt^2) + k^2(d^4z/dx^4) = 0$, which is, as well as I can remember (for I do not have the book at hand), the equation given by Euler for a vibrating elastic lamina. Your equation does not give this result. Your error seems to arise from the manner with which you tried to deduce the equation of a vibrating surface from the equation of a simple lamina; you became confused with the double integrals. They are nowhere amenable to the substitutions you have made. To obtain the equation of the surface one must follow the method indicated by Lagrange in his new edition, page 148, adjusting the suitable term to represent the force due to elasticity. There are also various other difficulties which have not yet been clarified, and there are even objections against the analysis of the article I cite.

    M. Biot, who has heard of your memoir, claims to have found the correct equation of the vibrating elastic surface. He has sent me an equation that he says he showed to M. de Laplace a long time ago, which is not the same as that which M. de Lagrange found from your hypothesis.

    I am not opposed to efforts that are praiseworthy in themselves, though they do not give the result I would have liked; but this is a further reason to remain incognito, and I promise on my part to preserve the deepest silence.

    I imagine that the same problem will be posed after a suitable time; therefore hope is not lost. On the contrary, one may dream more than ever of carrying away the palm.

    Accept, Mademoiselle, the affectionate sentiments of your devoted servant.

Legendre[19]

Sophie Germain's memoir did not deserve the prize, but it does merit some examination here. Basically her approach to the problem was through analogy, an often fruitful way of searching for truth, but not sufficiently trustworthy to guarantee valid results. Thus, starting from Euler's equation for the beam, she presented the following equation to represent equilibrium for any element of the plate:[20]

$$\int dz dy \int P ds + \int dz dx \int Q ds - 2\int dx dy \int R ds = V((1/r) + (1/r')).$$

She knew that this equation, however analogous to Euler's, wanted some justification. She tried to provide it in the supplement that she sent to Legendre after the competition deadline. Here she forced a derivation for this equation through a generalization, itself faulty, of Lagrange's treatment of a beam as it had appeared in the 1811 edition of *Mécanique Analytique.* Her technique was very much one of mathematical manipulation at the expense of physical understanding: she did not grasp the directional properties of force and moment; hence her attempt to derive an appropriate analysis of the two-dimensional elastic surface from the one-dimensional case was simply invalid.

In any case, this attempted derivation followed the construction of the equation rather than accounted for it. The equation itself, the basis for her entire effort, is in fact meaningless. It was an unsuccessful attempt to account for the equilibrium of externally applied forces (P, Q, and R) with internal forces generated by the resistance of the plate to deformation. This deformation was measured by the two principal radii of curvature of the deformed plate, r and r'.

Proceeding from this supposed equilibrium relationship, Sophie Germain made several assumptions that implied small displacements and rotations of the plate. She then differentiated four times with respect to x and y and obtained a sixth-order differential equation. Then, admitting only 'regular' solutions, by assuming that behavior in time would be harmonic, thereby eliminating the dependence on time, she deduced the following relationship:

$$z(x,y) = \frac{f^6}{2} \left(\frac{d^6z}{dx^4dy^2} + \frac{d^6z}{dy^4dx^2} \right).$$

This, then, was her equation, solutions of which she claimed should give the frequencies and vibration modes of an elastic surface. She continued by finding some special solutions based on Euler's model, showing some insight into formulating boundary conditions relying on considerations of symmetry.

All in all, Sophie Germain's memoir reveals brashness, speculation, and awkwardness. Basically, she was wrong, as Legendre's letter indicated. But Legendre also reported on an equation that Lagrange had derived from her work – an equation that displays a correct differential relationship – and, when supplemented with appropriate boundary conditions, provides today the basis for analysing both the static and dynamic behavior of plates. Lagrange had not found this equation independently of Sophie Germain's work, but had derived it through her memoir in a manner we can only surmise.

In Sophie Germain's basic equilibrium equation, the right-hand side consists of the expression $V((1/r) + (1/r'))$, which is meant to represent a plate's resistance to deformation due to its elasticity. The constant, V, represents the elastic nature of the material; the bracketed term, $((1/r) + (1/r'))$, represents the mean curvature of the plate. Thus she took the "elastic force" or "moment" (she uses both terms) to be proportional to the mean curvature of the surface. Just how she arrived at this relationship is unclear. In a footnote to her presentation of her basic equilibrium equation (see Figure V.4) she wrote:

On voit que je prends $V((1/r) + (1/r'))$ pour le moment de cette force. Il serait trop long de discuter ici les considérations que m'ont déterminé à choisir cette fonction des rayons osculateurs; mais au reste, il est aisé de sentir que lors même que l'expression du moment de l'élasticité devrait contenir d'autres fonctions de ces rayons osculateurs que celle que j'ai adoptée, tels que les produits ou les puissances des mêmes quantités, le resultat applicable à la question ne serait pas changé puisque comme on le verra par la suit $(1/r)\cdot(1/r')$, par example, pourra toujours être négligé vis à vis de $(1/r) + (1/r')^2$[21].

On the other hand, in the supplement she had stated that this relationship is a hypothesis drawn from analogy to Euler's expression for the elastic resistance of the beam.

Now, to Lagrange an expression for elastic force or moment meant something quite different from what it meant to Euler, and hence to Sophie Germain, although it is also unclear whether she fully understood Euler's conception. But whatever it meant to Sophie Germain, to Lagrange elastic moment was a scalar quantity that could be subjected to the methods of his *Mécanique Analytique*. He applied the variational method to her hypothetical expression and thereby derived from it the relationship now recognized as correct. He communicated his results to the other members of the commission in this way:

Note communicated to the Commissioners for the prize of elastic surface (December, 1811)

The fundamental equation for the motion of vibrating surfaces does not appear to me to be correct, and the way in which an attempt has been made to deduce this equation in passing from an elastic lamina as a line to a surface appears little justified. When z is small, the equation reduces to:

$$\frac{d^2z}{dt^2} + gEbc \left(\frac{d^6z}{dx^4dy^2} + \frac{d^6z}{dy^4dx^2}\right) = 0.$$

But in adopting, as has the author, $(1/r) + (1/r')$, for the measure of the curvature of the surface, which the elasticity tends to diminish, and to which one assumes it proportional, I find, in the case when z is very small, an equation of the form:

$$\frac{\mathrm{d}^2z}{\mathrm{d}t^2} + k^2\left(\frac{\mathrm{d}^4z}{\mathrm{d}x^4} + \frac{\mathrm{d}^4z}{\mathrm{d}x^2\mathrm{d}y^2} + \frac{\mathrm{d}^4z}{\mathrm{d}y^4}\right) = 0$$

which is very different from the preceding.[22]

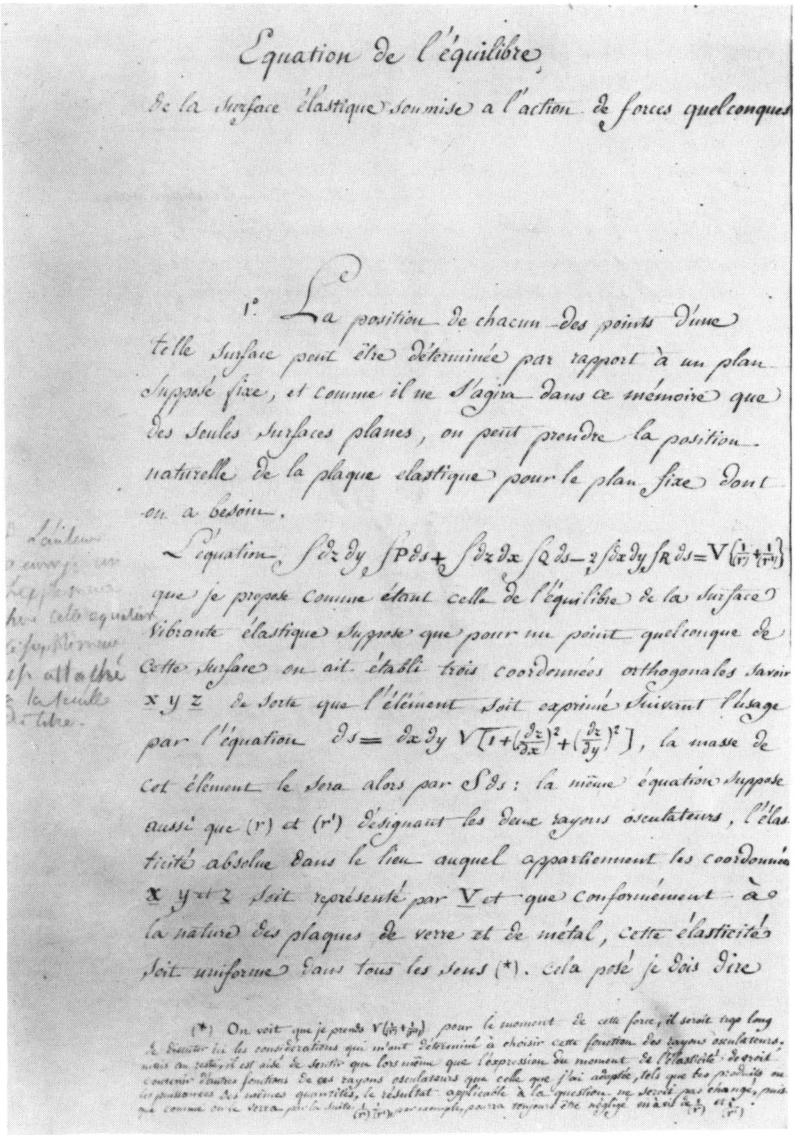

Fig. V.4   A first page from Sophie Germain's first entry.

Legendre then sent Lagrange's equation to Sophie Germain in his December letter, even indicating how she should proceed in order to arrive at Lagrange's result: ". . . follow Lagrange's method in the new edition (*Mécanique Analytique*), page 148, adjusting the appropriate term representing the force due to elasticity." But Sophie Germain was incapable of taking advantage of this advice. At least in 1811 she was unable to grasp fully the methods of the *Mécanique Analytique*. In her response to Legendre (or, more precisely, a draft of her response) she appeared a bit mystified and confused about what has happened.

Dec., 1811

I am not so surprised by the results you have informed me of since I had little confidence in my work. I was carried away by an analogy that seemed striking, but which I was not fully able to comprehend. I am most obliged to you for the care that you have taken to obtain a judgment for me and for enlightening me on the errors that I have made. What surprises me most is the equation that you indicated was obtained by Lagrange. If I did not know from whence it had come and, consequently, of the respect that it merits, I would have believed at first glance that it was formed from the sum of the two equations:

$$\frac{d^2z}{dt^2} + 2k^2 \left(\frac{d^4z}{dx^4} + \frac{d^4z}{dx^2dy^2} \right) = 0,$$

$$\frac{d^2z}{dt^2} + 2k^2 \left(\frac{d^4z}{dy^4} + \frac{d^4z}{dx^2dy^2} \right) = 0,$$

but I realize that I am in error.

If I was led astray by the double integrals, I ought not to be surprised that the equation obtained by Lagrange only attains to the fourth degree. I am convinced of the reality of my error, since it has been verified by judges whose opinions I respect, but I do not know in what it consists. If the equation

$$\int dxdy \int xdm + \int dxdy \int ydm - 2\int dxdy \int zdm = 2\int dzdxdy$$

is really deduced as I have tried to establish in the note, from the formulas of the *Mécanique Analytique* (second edition), does not it follow from differentiation four times, twice with respect to z and twice with respect to y? Is it not at least incontestable that in making an abstraction of the second member that contains the term due to the elasticity, and taking $x = y$, one obtains by this procedure of successive differentiations the equation:

$$R \left(\frac{d^2z}{dx^2} + \frac{d^2z}{dy^2} \right) = 2z$$

which is basically the same as that given by M. de Lagrange himself in the second section, and which belongs to the equilibrium of the surface pulled by equal forces along x and y? How can he thus hold that this same procedure is erroneous, when he applies it to the term depending on the elasticity?

You have informed me that the equation that M. de Lagrange has found in proceeding from my hypothesis includes the equation of the elastic lamina, whereas [the equation] I have deduced from the same supposition does not enjoy the same advantage. I too considered this same objection, but I believed that I had answered it in observing that for a quantity that is constant for the lamina, one would satisfy the equation by the supposition d(constant) = 0, such that the factor that composes the equation of the surface was not necessarily equal to zero when it concerns the lamina. Furthermore, I believed that the formula, when considered before differentiation, included the case of the lamina and it is this that I had the intention of proving at the end of the supplement. . . .

You have revealed to me a blemish in my work. I felt little prepared for this and since I have deceived myself to such an extent, to the same extent I am justified in distrusting my abilities. However, I like to arrive at the truth in order to remain certain of knowledge, even when it is not favorable to me. I have, moreover, happiness in having obtained, on this occasion, a new proof of the interest you deign to accord me. I sincerely assure you that your benevolence encourages me and that I value, as I should, the care that you take in order to hide me from the ridicule attached to my results. I pray you, Monsieur, to accept this expression of my recognition and to believe in the respect and admiration that I have always had for you, giving an infinite value to your attention.[23]

As Legendre had foreseen, the contest was extended. According to the announcement,

The Class, which has only received a single memoir, has thought that the time that they had accorded has not sufficed to permit the establishment, development, and confirmation by sufficient proofs of so difficult a theory. It has thus judged it appropriate to propose anew the same problem, in the same terms and same conditions. . . .

Efforts will only be received prior to the first of October, 1813; this condition will be rigorously observed.[24]

Sophie Germain had another year and a half to work on her analysis. The announcement of the competition required from her both a correct derivation of the plate equation proceeding from the supposition that the elastic force is some function of the principal curvatures (for example, proportional to $(1/r) + (1/r')$) and a justification of that particular functional relationship founded on an understanding of basic physical principles and processes. Her inability to master the variational techniques found in Lagrange's *Mécanique Analytique* accounts for her difficulty in meeting the first requirement. Determining the way in which the elastic force depends on the principal curvatures was a much more difficult problem. Lagrange had not succeeded in this, if indeed he tried. Both Poisson and Navier, as we shall see, floundered on this point. To Sophie Germain it was not just a difficult (if not impossible) problem, but also an ill-defined one.[25]

She knew, however, what was at issue in the mathematical derivation.

Legendre had given her the correct equation provided by Lagrange, and had indicated how she should go about deriving it. She put considerable effort into doing this, but with little success, since she lacked the mathematical competence.

She seems to have worked on her second memoir quite privately. At least there is no correspondence to suggest that she was in touch with any one else. Of course this is simply conjecture: the correspondence that remains consists of either letters addressed to her or drafts of letters that she sent (the recipients of her letters, other than Gauss, having not kept them). Still, the lack of letters suggests that she asked no further help, and the clumsiness of her final derivation suggests that she received none.

During the year and a half that she was working, however, she does seem to have gone through some personal changes. The devotion of time and energy, combined with some successes in predicting Chladni's patterns through her analysis, gave her increasing faith in herself and her accomplishments. Whatever errors she might make in working out details, she became convinced that her basic approach to the problem and the hypothesis she was working with were correct. At the same time, isolated as she was, she began to doubt the probity of the judges or their ability to judge well. She feared that Lagrange's statement on the difficulty of the problem would prejudice others' minds against recognizing a solution when one was found. Lagrange himself, who had at least worked with her hypothesis, would have been the one most likely to accept and appreciate her work, but he had died on the tenth of April, 1813.

As the time to submit her entry approached, Sophie Germain felt increasing confidence in herself and decreasing confidence in her judges. Some time near the October deadline she wrote the following letter, preserved in her draft without salutation. The tone of the letter suggests that she was not writing to Legendre. On the other hand, a certain distance may have come between them since the first contest, possibly because, as we have seen, he was not interested in helping her further with the problem. Such a refusal would also account for her comment that she did not expect him to be a judge. There is no way of being certain to whom she turned. But out of her isolation, her loneliness combined with determination and pride, she addressed someone of note and influence:

(circa October, 1813)

I enjoin your probation of memoir No. 1 carrying this epigram:

> But by far the greatest obstacle to the progress of science and to the undertaking of new tasks and provinces therein is found in this: that men despair and think things impossible.

If I had found the occasion, I would have consulted you before adopting this quotation, since it has an air of pretentiousness, which hardly suits me, having so many reasons to mistrust my own skills and, indeed, not seeing any strong objections to my theory other than the improbability of having it meet with justice. I fear, however, the influence of opinion that M. Lagrange expressed. Without doubt, the problem has been abandoned only because this grand geometer judged it difficult. Possibly this same prejudgment will mean a condemnation of my work without a reflective examination and it is this that has led me to place at the head of my memoir a quotation that seems appropriate to my thoughts in this regard. Moreover, I rely on the value of your support rather than on the influence of this philosophical thought of Bacon's, and although the subject matter of the memoir does not allow me to hope to have you as a judge, I approach you in order to ensure that the commissioners take the pain of understanding my long and tiresome work.

I have faith in the soundness of the theory that is its object. I have examined it many times. I have compared it to the results of M. Chladni's experiments and, moreover, I have not concealed the objection to it that one can draw from the difference between ratios (of frequencies) deduced from this theory and the ratios drawn from experiment. But this difference only appears in certain cases, namely those that are associated with the particular integral that leads to an equation of the same form as that for the vibrating string. It is hardly significant in other cases, and is evident only in the comparison of tones that are associated with the shapes included in this integral with tones that accompany the shapes that result from another particular integral. In fact, the tones given by each of these integrals, taken in the same circumstances, give, in the comparison among themselves, ratios conforming to experiment.

With regard to the mode shapes, I have explained them in a great number of ways that I believe satisfying. Thus, I think that my theory is supported by sufficient proofs and that it is more advanced, even without the comparison of ratios to experiment, than the theory of stretched surfaces, about which no one has expressed any doubts.

But even if I am mistaken in the principal objects of my investigation, there still remain some sections of my memoir that, perhaps, will not be unworthy of the attention of the Class. . . .

Despite all the reasons that I see in favor of my ideas, I have so little confidence in my judgment that I still doubt their value. How much I regret the advice of M. Lagrange! Even if he had condemned me, he would have at least have noted that that which is independent of the principal theory merits attention. If he had approved of this, I would have had proof that my work, as imperfect as it is, would have furnished him with the occasion of seeing the improvement of a theory in which I take a strong interest, independent of analysis, my proper love. Moreover, the notion that the problem is difficult, which perhaps might prevent one from devoting any effort to the examination of a memoir condemned once before, would not then arouse my fears. Similar regrets are also as natural as they are superfluous; they render your protection most necessary and I reclaim along with your confidence the interest, which ought to inspire me, and with which you have always honored me.

Perhaps you will pardon the length of my plea, which surely merits that I make you an excuse. I ask of you, Monsieur, to agree to the sentiments and admiration and respect of your servant.[26]

Her memoir was long, running to a hundred pages. Once again it was the only one submitted to the competition and it was received by the First Class on the twenty-first of September, 1813. Perhaps if Poisson had not been elected to the Physics Section of the First Class to the vacancy created by the unexpected death of Malus, there would have been two entries. His election in March, 1812, however, precluded his entering the competition; his own analysis of the plate vibration phenomenon did not appear until 1814. Before that, however, Poisson was to act as judge rather than competitor. He, along with Laplace, Lacroix, Carnot, and once again Legendre, had the task of evaluating Sophie Germain's second entry.

In the opening section of her memoir she presented a derivation of the differential equation for the plate. To do this she first had to establish an expression for

la somme des moments, des forces d'élasticité qui agissent dans toute l'étendue de la surface.[27]

This expression had then to be operated on with the variational method, in order to derive the plate equation. Sophie Germain knew what her goal was as well as the point from which she should start. She therefore did obtain the correct equation, namely, that produced by Lagrange. Yet she presented no justification for her expression for the elastic force – other than that it was "the most convenient" – and her method was so awkward and so full of errors of both commission and omission, that her derivation was totally unconvincing to the judges; nor was there any reason that they should have been convinced.

After this initial, and somewhat disastrous analysis, she went on to obtain particular integrals to the plate equation. Here she operated in more familiar territory and generated four forms of solutions.

To complete the solution, to determine particular mode shapes and frequencies, she had to establish appropriate boundary conditions. Her preparation of these conditions proceeded from a lengthy discussion of Euler's investigation of the vibration of a beam, in the course of which she emphasized the difference between a node and the conditions that hold at a point of support. She then obtained from her variational derivation, despite its clumsiness, some boundary conditions applicable to the plate problem.

With this handful of integral and boundary conditions, she turned to the task of drawing out specific solutions to compare with Chladni's results. She focused first on patterns observed in vibrating square plates, and showed how the results of her analysis predicted the disposition of nodal lines and the frequency ratios corresponding to different mode shapes. In another section she treated rectangular plates. Here again she was able to argue that selected solutions predicted certain experimentally observed mode shapes and frequency ratios. This correspondence between experimental shapes and mathematical predictions impressed the judges more favorably than her derivation.

The first indication of how her memoir was received came again from Legendre, in a letter written early in December, replying evidently to her request once more to add something to the submitted memoir. Here, Legendre expressed not only judgment but irritation as well. We have no evidence whether, or to what extent, she had made many requests of him earlier, or embarrassed him before his colleagues. Possibly his championing of her cause during the first competition had raised questions or eyebrows when it came to his election to the second commission of judges for the second contest. In any case, we hear him addressing his *protégée* in a new tone.

Mademoiselle,   Dec. 4, 1813

I do not understand the analysis you send me at all; there is certainly an error in the writing or the reasoning, and I am led to believe that you do not have a very clear idea of the operations on double integrals in the calculus of variations. Your explanation of the four points does not satisfy me any more. Lagrange was correct in considering two consecutive elements on the elastic curve and measuring the elasticity by the angle made by these two elements. There are no analogous elements in surfaces, or at least those we have considered are not an analogy. An element of the surface has a projection $dxdy$; the element [after deformation] has a projection of $(dx + ddx)(dy + ddy)$. These two projections are two different squares. In addition, the nature of planes does not accommodate these projections, since a plane does not pass through four points. There is a great lack of clarity in all of this.

I will not try to point out to you all the difficulties in a matter that I have not especially studied and that does not attract me; therefore it is useless to offer to meet with you and discuss them. Besides, the thing is over with; there is nothing further to change in the memoir, and with all my good will I can do nothing.

However, it seems to be recognized that your equation is truly that of the vibrating surface. Putting the analysis aside, the rest, concerning the explanation of the phenomena, may be good. If the commission of the Institute were of this opinion, you might at least receive honorable mention; but I hope that the incorrect analysis will not harm the rest of the memoir and the parts of it that are correct.

In any case, there is the possibility of having your research published, reestablishing the correct analysis or downplaying it, and your work will bring you honor. This was perhaps the proper thing to have done in the first place. But I promise you always the deepest secrecy, and, if you have not committed some other indiscretion, it will be as though the thing were null and void.

Accept, I beg, my respects and my complete devotion.

Legendre[28]

As Legendre expected, Sophie Germain's memoir was awarded an honorable mention. Yet this did not lay the matter to rest either for her or any of the others engaged by the problem. The competition itself was again prolonged, with the following announcement:

. . . the analysis that the author of this piece has employed in order to come to his fundamental equation has been judged completely incorrect, and this same equation appears not to result in any way from the method of analysis. But the part of this memoir that includes the comparison of theory with the experiments of M. Chladni being made with care and leading, in general, to satisfactory results, the class has reasoned that this piece merits an honorable mention.

The class proposes anew the same problem in the same terms and conditions. . . .

Works will only be received prior to 1 October, 1815; this condition is firm.[29]

CHAPTER SIX

THE MOLECULAR MENTALITY

Entries for the third competition were due at the beginning of October, 1815. During the session of the First Class on 1 August, 1814, however, Poisson began reading a memoir in the following way:

> The efforts of mathematicians of the last century have brought the science of mechanics to a level of perfection that has given rise to the opinion that this science is completed and that there remain only some difficulties of the integral calculus to overcome in each particular problem. This is not the case, however; mechanics still presents several important problems that have not yet yielded to calculation; the theory of elastic surfaces, which I propose to consider in this memoir, offers a remarkable example. The differential equations of these surfaces in static equilibrium, and more importantly, those of their movement, are not yet known.[1]

Once before Poisson had presented a memoir to the First Class on a topic that was a subject of a public competition. That was in 1812, prior to his election to the First Class, when his memoir concerning the distribution of electric charge on objects had occasioned the retirement of a prize. Now, however, circumstances were different. Poisson had become a member of the Class, and in 1813 he had been one of the judges of the competition for the *prix extraordinaire*. It was not the custom for members of the First Class to compete for prizes they themselves had established. Surely Poisson's reading of memoir addressing the analysis of vibrating elastic surfaces could be considered a breach of proper conduct.

Legendre felt it to be so. He actually interrupted the reading to object. The discussion that followed was evidently inconclusive and unsatisfying (the minutes of the meeting do not give a detailed account of it) since a committee was formed to consider the matter further and Poisson was allowed to continue his reading. The committee, however, was never mentioned again in the minutes of the proceedings of the First Class.[2]

What Poisson claimed in his memoir was a rigorous derivation of a differential equation for vibrating elastic surfaces. It was for just such a lack of rigorous derivation that the judges had faulted Sophie Germain. Not surprisingly, the equation that Poisson derived was the one that Lagrange had produced and that Sophie Germain had tried to derive.

65

When Lagrange had first derived that equation and put it in his note to the judges of the first contest, it had little or no stature. Lagrange simply presented it as the correct equation *if* Sophie Germain's hypothesis was correct. At the same time he showed that, even granting her hypothesis, her mathematics was in error. By the end of the second contest, however, she had demonstrated that this equation worked to predict vibration patterns in a number of special cases. That demonstration made the equation far more interesting than it had previously been: rather than simply implicit in a random hypothesis that might or might not have anything to do with physical reality, it showed promise of being a correct equation. So Poisson reasonably, if somewhat unethically (since it had not been made public), set about trying to derive this equation rigorously.

Poisson pursued this task far differently than Lagrange or Sophie Germain had. The questions that concerned him focused on how to derive this equation from the mathematical model in which he believed: the molecular hypothesis that was described and established by Laplace.

Far removed from the scientific certainties of Napoleonic France and Laplace's hegemony in scientific affairs, one can now speak of that molecular hypothesis as both a hypothesis and an error. At the time, however, it was a basis for discovery and analysis. Poisson could, and did, claim that he was working without a hypothesis, suggesting that he was working with the facts of reality itself, not merely with an abstract conceptual scheme. This way of perceiving his work and thought reflects not a molecular hypothesis, but a pervasive molecular mentality. It is this with which Sophie Germain had to contend and from which she was intellectually free.

Laplace's approach to the analysis of physical phenomena was based on the work of Isaac Newton. Two centuries earlier, Newton had introduced and employed the concept of gravitational force, a force inherent in any matter, with matter alone all that is required to affect its action. It acted through the void of celestial space, yet remained undisturbed by the interpositioning of more of the same or other matter. Any two particles at a distance apart attract each other with a force proportional to the product of their masses and inversely proportional to the square of that distance. With this 'inverse square law,' Newton provided a foundation for the explanation of the motion of all celestial bodies. The motion of terrestrial objects also fell under its aegis: the descent of an apple was determined by the same law that explained the motion of the Moon about the Earth.

Laplace expressed his appreciation for Newton's achievement in the first edition (1796) of his *Exposition du Système du Monde:*

One can increase a theory's probability either by diminishing the number of hypotheses upon which it is based or by augmenting the number of phenomena it explains. The principle of gravity has brought these two advantages to the theory of the movement of the Earth. In order to explain the apparent movement of the stars, Copernicus endowed the Earth with three distinct movements: one about the Sun; another of revolution by itself; finally, a third movement of its poles about those of the ecliptic. The principle of gravity makes all three depend on a single motion imparted to the earth, along a direction that never passes through its center of gravity. By virtue of this movement, it turns around the Sun and on itself. It has taken on a flattened shape at its poles and the action of the Sun and Moon on this shape makes the axis of the Earth move slowly about the poles of the ecliptic. The discovery of this principle has thus reduced the assumptions upon which Copernicus founded his theory to the smallest possible number. Furthermore, it has the advantage of relating this theory to all astronomical phenomena. Without it, the elliptic orbits of the planets, the laws that the planets and comets obey in their movement around the Sun, their secular and periodic inequalities, the numerous anomalies of the Moon and the satellites of Jupiter, the precession of the equinoxes, the nutation of the Earth's axis, the movements of the Moon's axis, finally the flux and reflux of the sea, would only be the results of observations, isolated from one another. It is truly admirable how all these phenomena that seem, at first glance, so disparate, flow from the same law that relates them to the movement of the earth such that, once this motion is admitted, one is led by a path of mathematical reasoning to these phenomena. Each of them thus furnishes a proof of its existence. And if one considers that there is not a single one that could not be led back to the law of gravitation, that this law determines the position and movements of celestial bodies at each instant in all their travels with the greatest exactitude, there is no fear that it could be devalued by some phenomenon hitherto unobserved. Finally, the plant Uranus and its newly discovered satellites obey and confirm it. It is impossible to reject it in the face of this collection of proofs and to deny that nothing is better demonstrated in natural philosophy than the motion of the Earth and the principle of universal gravitation, proportional to mass and reciprocally proportional to the square of the distances.[3]

Laplace did not apply his expertise to problems in celestial mechanics alone, "augmenting the number of phenomena" explained by universal gravitation. In the eighteenth century many who were committed to the pursuit of scientific knowledge, enlightened by Newton's achievement, had labored to develop corpuscular-force theories that would account for a variety of physical and chemical phenomena. Indeed, Newton had argued that certain of these could be explained through forces acting reciprocally between elements of matter where these forces do not necessarily obey the inverse square rule.[4] Laplace joined in this endeavor and attempted to discover what these corpuscular forces might be, what rules they would follow, and how they could be used to predict and explain diverse terrestrial phenomena, just as gravitation could predict and explain such varied celestial phenomena.

This project engaged Laplace with varying intensity throughout his entire career. It was a developing scheme, conditioned by mathematical technique, elaborated when necessary to encompass new phenomena, but rooted in one economical, general principle: the existence of "sensible forces at insensible distances" acting between molecules of substance, just as gravitation is the prevailing force acting between large bodies at measurable distances.[5] In 1796 he had written:

The attractive force of universal gravitation vanishes between bodies of extremely small magnitude; it reappears in their elements under an infinity of different forms. The solidity of bodies, their crystallization, the refraction of light, the raising and lowering of fluids in capillary tubes and, in general, all chemical combinations are the result of attractive forces, the knowledge of which is one of the principal objectives of physics.[6]

These forces of attraction, Laplace went on to suggest, may indeed be due to gravitation, but the resultant force law, at insensible distances from the molecule, might be modified by the shape of the individual molecule, just as the resultant force field due to gravitation in the immediate vicinity of an oblate planet does not follow the inverse square law:

These affinities would depend then on the shape of individual molecules and we could, based on the variety of these shapes, explain all the varieties of attractive forces, and in that way, reduce all phenomena of physics and astronomy to a single general law.[7]

He proceeded to argue, however, that it is impossible to determine these shapes, thus rendering this line of research useless. Unless these shapes were known, there was no way of deducing how the resultant force of affinity varied with distance. Laplace noted that:

Several geometers, in order to provide a rational basis for these affinities, have added to the inverse square law of attraction new terms that are only sensible at very small distances; but these terms could represent many different forces. In toying, moreover, with the shape of molecules, they only complicate the explanation of phenomena.[8]

Although Laplace himself preferred one law of affinity – gravitation modified by the shape of molecules – rather than many force laws, conjured up to save the appearances of a variety of phenomena, the forementioned problems in determining those shapes left him in a bit of a quandary:

In the midst of these uncertainties, the wisest part is to attempt to determine by numerous experiments the laws of affinities. . . . Some experiments already made in this manner give reason to hope that one day these laws will be perfectly known; then in applying the calculus, one could elevate the physics of terrestial bodies to a level of perfection that the discovery of universal gravitation has given to celestial physics.[9]

But whatever force laws there might be, they were not to be revealed by experiment.

During the next few years, however, Laplace experienced a breakthrough. He discovered that it was not necessary to specify explicitly the laws of attraction operating among molecules in order to accomplish meaningful analysis of some phenomena, i.e., one need not know the way a force of affinity varied with distance. All that was required was that these forces exist and that they only be "sensible at insensible distances." The significance of this expression becomes clear through a consideration of how it was used by Laplace in the analysis of specific phenomena.

Laplace's search for a molecular parallel to Newtonian gravitation and his now rather obscure stipulation that it consist of forces sensible at insensible distances shaped an intellectual milieu that fostered much of the serious work in physics in Paris during the first decades of the nineteenth century. This molecular mentality pervaded the scientific activity of the Society of Arceuil, a small select group of *savants* organized and guided by Laplace and Berthollet and whose members included Biot, Thenard, Gay-Lussac, Malus, Arago, and eventually Poisson. Their efforts, primarily experimental in nature, were directed to the exposition of a variety of phenomena: the behavior of gases, chemical combinations, the physics of light, the velocity of sound – endeavors all influenced by the principle of sensible forces at insensible distances.[10]

Even after the dissolution of this society, Laplace's promotion and often successful application of molecular analysis continued to stimulate and provide a basis for scientific effort. Joining the molecular force principle to a material theory of heat proved a profitable exercise for Poisson as well as Laplace in the analysis of the behavior of gases. In a treatise read before the Academy in September of 1821, 'Sur l'attraction des corps sphériques, et sur la répulsion des fluides élastiques,' Laplace assumed that:

The molecules of a gas are at such distances that their mutual attraction is insensible. . . . I assume also that these molecules retain heat [*chaleur*] by means of their attraction, and that their mutual repulsion is due to the repulsion of the molecules of heat, repulsion whose extent I assume is over an insensible sphere of activity.[11]

This way of perceiving the action of heat served Laplace in the same memoir as a basis for an analysis leading to an expression for the velocity of sound and its explicit dependence upon the specific heat coefficients of a gas – a relationship first announced, but without proof, in 1816.[12]

Laplace's devotion to his molecular principle is expressed in the following terms at the conclusion of this treatise:

I hope that this new extension of the theory of these forces will interest geometers. Nearly all phenomena of terrestrial nature depend on it, just as celestial phenomena depend on universal gravitation. It appears to me that this theory now ought to be the principal object of research in mathematical philosophy; it seems useful to me to introduce it into the demonstrations of mechanics, abandoning abstract considerations of massless flexible or inflexible lines. Several attempts have made me see that, by approaching nature in this way, one can give to these demonstrations more simplicity and much more clarity than by the methods followed up to the present.[13]

A measure of Poisson's allegiance to Laplace and the molecular conceptual scheme is revealed in the following quote. It also displays Poisson's less than full confidence in the methods espoused by Lagrange in the *Analytique Méchanique*, methods that served as a basis for Sophie Germain's work.

Let us add that it would be desirable for geometers to reconsider the principal questions of mechanics from this physical point of view, which conforms to nature. It has been necessary to treat them in quite an abstract manner, in order to discover the general laws of equilibrium and movement. In this kind of generalization and abstraction, Lagrange went as far as one could conceive when he replaced the physical links of bodies by equations between the coordinates of their different points. It is this that is the essence of his *Mécanique Analytique*; but aside from this admirable conception, we can now elevate the *Mécanique physique*, of which the unique principle will be to reduce everything to molecular actions that transmit, from one point to another, the action of the given forces and that are the intermediary of their equilibrium. In this way, one would have to make no special hypothesis when one wished to apply the general rules to the mechanics of particular problems.[14]

In 1814 Poisson was very much under the influence of the molecular mentality. If, as seems most likely, he had been working on elasticity for several years, he had almost certainly also been approaching the problem from a molecular view. This inference is based not only on the intellectual milieu in which he worked, but also on his patron's explicit guidelines. In 1809, the year of Chladni's visit to Paris, Laplace had written:

In general, all attractive and repulsive forces in nature reduce, in the final analysis, to similar forces acting between one molecule and another. It is this that I have shown in my 'Theory of Capillary Action,' that the attractions and repulsions of small particles floating on a liquid, and, in general, all capillary phenomena, depend on the attraction of one molecule for another – attractions that are sensible only at imperceptible distances. Similarly, we have attempted to reduce electric and magnetic phenomena to intermolecular action. Furthermore, we can reduce those phenomena presented by elastic bodies. In order to determine the equilibrium and motion of a naturally straight, elastic lamina, bent along some arbitrary curve, it has been assumed that at each point its spring is inversely proportional to the radius of curvature. But this law is only secondary and derives from the attractive and repulsive action between molecules.[15]

Then, at the August 1, 1814, meeting of the First Class, when Poisson presented his memoir on elastic surfaces, he expressed the Laplacian molecular principle in the following terms:

Whatever be the cause of the elasticity of bodies, it is certain that it consists in a tendency of their molecules to mutually repulse each other, and one can attribte this to a repulsive force acting between them according to a certain function of distance. Moreover, it is natural to think that this force, as with all other molecular actions, is sensible only up to imperceptible distances. The function that expresses this law ought thus to be regarded as zero whenever the variable that represents distance is no longer extremely small: now one knows that similar functions vanish, in general, in the calculus and only leave, in the definitive results, total integrals or arbitrary constants that are obtained from observation. This, in fact, is what happens in the theory of refractions and still further in the theory of capillary action, one of the most beautiful applications of analysis to physics that is due to mathematics. The same happens in the present problem and it is this that has permitted the expression of forces describing the elasticity of a surface in terms of quantities uniquely dependent on its changing shape, such as its principal radii of curvature and their partial derivatives.[16]

This bald statement merits some examination and application since understanding the conceptual schemes of another age is not a trivial problem, especially when the later developments have shown the premises of the earlier time to be wrong and have rendered once confidently embraced assumptions questionable, if not absurd. It would be useful first to examine the theory of refraction, which Poisson calls upon as a parallel to the work he is about to present.

In the first edition (1796) of the *Système du Monde*, Laplace had referred to both refraction of light and capillary action as examples of terrestrial phenomena that needed analysis. In 1805 he presented an analysis of refraction of light within the molecular conceptual scheme that illuminates the phrase 'sensible forces at insensible distances.'

Consider a molecule, of light, impinging at some angle on a transparent surface; a surface which is also constituted of molecules.[17] Each molecule of substance acts on the molecule of light with an attractive force. To Laplace, this force is sensible, determinate of the molecule's motion, only at insensible distances, i.e., distances so small that he may advance the following argument:

In fact, the action of bodies on light being sensible only at very small distances, the parts of bodies a little distant from the molecule of light never have any sensible action on it and we can, in the calculation of the action of bodies, consider it an infinite solid terminated by an indefinite, plane surface in all directions. With this hypothesis, it is clear that the action of bodies on the molecule of light is perpendicular to its surface.

From this seemingly modest hypothesis Laplace proceeds to deduce Snell's Law in the form:

$$\sin\theta' = \frac{\sin\theta}{\sqrt{1 + 4\rho k/n^2}} \, ,$$

where θ is the angle of incidence the ray of light makes with the perpendicular to the surface before entering the body and θ' the angle the refracted ray makes with this same perpendicular. n is the speed of light at a sensible distance from the medium, ρ the density of the solid, and k a constant representing the attractive potential of the refracting medium. k depends on the law of force attracting molecule of light to molecule of substance, but only through an integral over an infinite spatial range:

$$k = \int_0^\infty \Pi(s) \cdot \mathrm{d}s.$$

The supposition 'sensible at insensible distances' permits these limits to be so defined. The result is that Laplace did not need to specify the exact nature of the force law, $\Pi(s)$, i.e., its explicit dependence on distance s. In fact it may be that

[This quantify $(4\rho k)$] is not the same for different transparent bodies; it is not proportional to their densities (ρ); it is possible that the function of distance that expresses their action on light is different for each body; it could also be the same for each body while it only differs in different bodies by the product of their densities multiplied by a different constant coefficient: different according to their nature. In both of these assumptions, the total action of bodies on light will be the same and, since in the integration it is only the total resultant of this action that matters, and one can employ the second supposition as the simplest.[18]

That the force law cannot be determined as a function of distance made no difference to Laplace's deduction of Snell's law. Different materials may indeed attract a molecule of light according to different functions of distance or, more simply, the dependence of force on distance may be one and the same for different materials, and the difference in the resultant effect of attraction accounted for by the scaling of force-intensity by a factor that does vary according to the material. The parameter k remains to be determined by experiment. This task is obviously less demanding than that of determining the variation of the sensible force at insensible distance with distance, a program of research originally suggested by Laplace.

Laplace's treatment of capillary phenomena was based on the same inter-molecular force mechanism, but required more elaborate analysis. In deriving Snell's law, Laplace was working toward a known end: the law had been derived earlier by other arguments. Laplace was simply deriving the known law within his molecular conceptual scheme (somewhat as Poisson was later deriving Lagrange's and Sophie Germain's equation within the same scheme). When, however, he came to the raising and lowering of fluids in capillary tubes, he was dealing with an unexplained phenomenon.[19] Not only did he explain the 'true cause' of the experimentally observed fact that the elevation or lowering of liquids in small diameter tubes is universely proportional to the tubes' diameters, but, in the course of his analysis, he derived the partial differential equation defining the shape of the liquid's free surface.[20] He then analyzed a variety of phenomena: the attraction and repulsion of small particles floating on the surface of liquids; the mutual attraction of two parallel plates partially submersed in a fluid; the suspension of particles at the surface of a liquid whose specific gravity is less than that of the particles. Indeed, he could conclude:

One of the greatest advantages of mathematical theories, and the most appropriate to establishing their certainty, consists in tying together phenomena that appear disparate by determining their interdependency, not.through vague consideration and conjecture, but by means of a rigorous calculus. Thus the law of gravitation relates the flux and reflux of the sea to the law of the elliptical movement of the planets. In the same way, the preceding theory makes the adhesion of discs to the surface of liquids depend on the adhesion of these same liquids in capillary tubes.[21]

It was this work by Laplace, summarized in the 1808 edition of his *Système du Monde*, that Poisson called upon as parallel to his work in elasticity.[22] His derivation of the plate equation proceeded from equilibrium considerations of an isolated molecule of the elastic surface. The deformation of the surface, its stretching and bending, changed the distances between molecules. The changed distances generated the force of elasticity that tended to return the surface to its original flat configuration. This derivation is a *tour de force*, requiring extensive expansions and truncation of series. As one reads through his demonstration it becomes increasingly difficult to associate his mathematical statements with the physical features of the phenomenon.

Finally his analysis resulted in the following equation:[23]

$$n^2\epsilon^2 \left[\frac{1 + q^2}{k} \frac{d^2P}{dx^2} - \frac{2pq}{k} \frac{d^2P}{dxdy} + \frac{1 + p^2}{k} \frac{d^2P}{dy^2} - p\frac{PdP}{dx} - q\frac{PdP}{dy} + \frac{kP}{2}(P^2 - 4Q) \right] = Z - pX - qY - kP\pi.$$

In this P and Q are explicit functions of the principal radii of curvature: $P = (1/r) + (1/r'); Q = 1/rr'$.

This is a frightening equation, fraught with nonlinearities. It is truly a wonder that it leads, after appropriate simplification (linearization), to the equation of motion for the vibrating plate.

From the viewpoint of a twentieth-century elastician, the use of this molecular mechanism as a basis for equilibrium considerations seems idiosyncratic, if not simply absurd.[24] It enabled Poisson to derive an expression that includes bending from a model that does not include bending, that is, the molecular model does not account for the redistribution of molecules over the thickness (their compression at the top of the surface and dilation at the bottom, causing the surface to bend upwards). Rather, he dealt with an abstract plane and the relationship of molecules to one another in that plane. There is no reason to think that Poisson could have derived the equation he did by the method he used unless he had known the equation he was attempting to derive. Of course he did know that equation, since he had been a judge for Sophie Germain's second memoir. He might have heard of the equation earlier, since Lagrange had derived it from Sophie Germain's first memoir, and clearly word of such things passes among scientists (as it passed to Sophie Germain herself). Yet before her second memoir that equation had not appeared particularly interesting: it was simply the equation Sophie Germain could have derived from her hypothesis had her mathematical understanding been good. It said nothing about the validity of the hypothesis. The second memoir, however, demonstrated the predictive powers of that equation in a number of special cases. The equation then merited serious attention. And Poisson, who valued Lagrange's methods rather little, could easily dismiss the hypothesis from which the equation derived, while accepting the equation as, simply by chance, correct. He then derived it from the molecular model, as he believed appropriate and true.

Certainly this is a normal mode for a mathematician to work in: in abstract mathematics, the derivation of a theorem oft-times, perhaps always, follows the discovery of the theorem. In this instance, however, it is clear that Poisson could not have derived that equation from the molecular model without knowing at the outset what his goal was. Further, Poisson was not about to employ the molecular approach to justify the hypothesis promoted by Sophie Germain, i.e., that the elastic force, in the Lagrangian sense, is proportional to the mean curvature. Only a mind unconstrained by the molecular mentality could come up with that hypothesis and Sophie Germain's mind was quite innocent of any molecular prejudices.

When Poisson presented his memoir on elastic surfaces to the First Class, he noted that the plate equation had first appeared, without proof, in the work of the anonymous author who had been awarded an honorable mention; but he did not restrict announcement of his work to the First Class in order to avoid influencing the competition still under way. Instead he offered a shortened version of his memoir to the *Bulletin des Sciences, par la Société Philomatique de Paris*. Since Poisson himself was the editor overseeing mathematical contributions to this journal, his memoir was readily accepted and appeared in print late in the year 1814.[25]

A report of Poisson's contribution also appeared in *Correspondance l'École Polytechnique*. The editor of this publication attempted to smooth out the apparent inconsistency and irregularity of these proceedings with the following footnote:

The Institute's class of mathematical and physical sciences has reset the contest on the theory of vibration of elastic surfaces for the year 1815. This extract of a memoir by Poisson on elastic surfaces will be very useful to those young geometers who compete for the prize; that is what has led me to insert it in our *Correspondance*, although it has already appeared in a *Bulletin, Société Philomatique*.[26]

Sophie Germain would not take the advice of the editor. Not only was she not a "young geometer" (she was thirty-nine at the time), but she knew perfectly well where the equation had come from and from what hypothesis it had been originally derived, even if she still lacked the insight that would enable her to make a properly rigorous analysis. She was unimpressed by Poisson's accomplishment, so she continued on her own path toward a prize that by now – however modestly she had begun – she felt was rightfully hers.

On the other hand, Poisson's work on this problem demonstrates not only the machinations of the powerful within a scientific elite, but the power of a dominant mind-set within such a group. The molecular mentality was an established view, a way of modeling, a framework for analysis promoted and employed to meaningful ends by a figure of authority in French science and his disciples and not without effect on others.[27] It shaped the criteria on which their judgments were based as well as Sophie Germain's efforts in search of scientific explanations and knowledge. It is only within the context of the molecular mentality that one can measure the meaning of expressions like 'without any hypothesis' and 'rigorous demonstration.' It was a paradigm, a way of behaving, of setting rules as well as defining problems.

CHAPTER SEVEN

AN AWARD WITH RESERVATIONS

After Poisson read his 1814 memoir he had every reason to expect the prize would be retired. It had been established with him in mind and he had finally realized Laplace's expectation of him. It had taken longer than anticipated and he had needed help, but still he had completed an analysis. The help had come first of all by chance from Lagrange. Sophie Germain had provided more help by demonstrating that Lagrange's equation, in fact, worked in a number of special cases, experimentally demonstrating its validity. Yet while Poisson could use this information to encourage and guide him in his own work, he did not have to think seriously of Sophie Germain as either critic or competitor. Apparently her mathematical acumen was not great enough to allow her to derive Lagrange's equation in any way. The honorable mention awarded her – by Poisson as well as the other judges – constituted acknowledgement of rather useful work by an inferior, rather than recognition of a peer.

Furthermore, the derivation that Poisson presented was totally satisfying to his contemporaries. Lagrange might have questioned Poisson's derivation from the molecular model, were he still alive. Biot, in a review of memoirs presented to the Class of Mathematical Sciences for the year 1814, wrote:

. . . we first consider mathematics, that powerful lever of human intelligence. In this genre, we find a memoir of M. Poisson on the acoustic vibrations of elastic surfaces – a remarkable memoir, due as much to the results that it contains as to the extreme difficulty of the subject. Mathematicians of the last century have been very fortunate; the heavens, which Newton opened to them, offered to their wisdom a vast domain for work and discovery. They have left to their successors few things to research; nothing, at least, which appears vitally important to the progress of astronomy and its application to geography and navigation. For us today it is necessary to search for another career in physical and chemical phenomena. Yet what a difference in complication and consequences! Celestial motion, as complicated as it appears, only depends on the reciprocal action of a small number of bodies located at great distances one from another, moving in the void with a remarkable regularity. If it has required so much effort to develop all these laws, how still much greater difficulty would one expect to experience in calculating the reciprocal actions of an infinity of particles, so close to each other that even their shape has a sensible influence on their effects! Such, however, is the nature of the problems that physics and chemistry present, since they always involve actions at small distances on bodies of sensible extent. The relevant problems in electricity, heat, and light are of this kind. The problem of elastic surfaces, which M. Poisson has resolved, is also of this kind; this essay ought to clarify what interest his solution should have, not only because of the immediate results that derive from it, but for bringing to perfection the analytic method employed.[1]

Despite all this, the prize was not retired. Sophie Germain, however, at first responded to Poisson's reading (any one of her friends in the Academy would have told her about it immediately) by stopping her work on the problem. How could she not? She found herself somewhat in the position of a mystery writer who has been working in almost total isolation on an intricate and fascinating plot that will make her name known as author. Suddenly a new book appears by an already famous writer with virtually the same plot, but far more brilliantly motivated and cogently argued. The novice author may suspect plagiarism, but is quite helpless. She has no choice but to give up the project.

Before this time, Sophie Germain had indeed been working in relative privacy, always hopeful of being able to present to the scientific community both a beautifully derived equation and the evidence that it was correct. Now the equation had been made public, and the only remaining problem for the competition was that of demonstration. She did not seriously have to fear that others might take advantage of the publication of the equation to enter the competition at this late date. No, there was scarcely any motivation for new work, either by a new person or by Sophie Germain.

In addition to these pressures to give up the competition, Sophie Germain would have had every reason to doubt impartial judgment if she clung to her own hypothesis; surely Poisson would be among the judges as now the ranking specialist in the field, and he had made his opinion clear. Nor was she about to change her way of attacking the problem. Of all people she was least likely to be impressed by Poisson, and she was certainly not a person to alter her work for the personal and transient glory of winning a prize.

On the contrary, Sophie Germain was stubborn and headstrong and, upon hearing of Poisson's memoir, no doubt angry and resentful as well. She might abandon her work in disgust and despair, but she would take the first reasonable opportunity to resume it and try thereby to claim respect and perhaps even recognition of her originality from the man who had ignored her work totally, save to use what he could of it for his own advantage.

The opportunity probably arose through Legendre, Sophie Germain's continuing friend and protector behind the scenes. His objection to Poisson's reading had resulted in the appointment of a commission to study the propriety of the case. This commission never made an official report, nor did Legendre press his complaints further; however, the prize was continued. In view of the politicking that so often took place, it seems most likely that an oral agreement was reached among Legendre, Poisson, Laplace, and any other interested parties that the contest would be continued, that the prize would be awarded to Sophie Germain if she could produce a memoir at all worthy, and that Legendre would drop his complaint. Legendre would have told Sophie Germain of this agreement in person; he was concerned enough about indiscretions to avoid committing it to writing.

Such an agreement was just the encouragement that Sophie Germain needed. Not only did it alleviate her fear of hostile judgment, but it also invited her to pursue her work with the full independence of her nature, despite Poisson's contribution. In her third entry, after showing that her hypothesis yielded the more general, nonlinear equation produced by Poisson as well as the equation obtained first by Lagrange, she wrote:

J'ai vivement regretté de ne pas connaitre le mémoire de Mr. Poisson. J'ai passé à en attendre la publication un temps qui m'eut été precieux. J'aurais même entièrement renoncé aux recherches que j'ai l'honneur de soumettre au jugement de la classe si je n'avais appris, . . . que l'équation obtenue dans une hypothèse differente de celle que j'avais proposée, résulterais également de cette dernière hypothèse. En effet je voyais chaque jour de nouvelles raisons de regarder mon hypothèse comme incontestable; et pourtant le respect dû à l'autorité de Mr. Poisson m'otais le courage de soumettre au calcul un principe que je ne prévoyais pas alors devoir être d'accord avec l'équation publiée par cet habile géometre.[2]

Here she is asserting that Poisson's analysis, based on a different hypothesis, increased her faith in her own work. This is not altogether wilful on her part. First of all, she did not have access to Poisson's full work to be convinced by it if it were convincing. Although she may have heard of his declaration that he worked from basic truths more solid than mere hypotheses, she had no reason to believe him, since she did not recognize his basic truths. She believed in Lagrange's equation that was derived from her hypothesis, even if she couldn't do the mathematics herself, and she accepted Legendre's implicit faith in that equation – as shown by his trying to direct her to the proper method for its derivation. Now Poisson's work increased her faith further, since he also accepted that equation. She was not apt, as a rationalist, to believe that the same correct equation would be derived from two noncommensurate theories; it seemed too obvious that a correct method was what gave correct results. Thus she could assume that Poisson was deriving what he knew and accepted, and this act served to validate the hypothesis from which the plate equation had originally sprung.

Sophie Germain's third memoir is about half the length of the second entry and departs from her previous work in significant ways. In the first place, she attempted to extend her research to include the vibration of initially-curved surfaces. This was an appropriate direction for her to move in but, even discounting the fact that her basic hypothesis was inadequate when generalized to apply to such surfaces, her abilities, particularly in employing variational methods, were not equal to the problems she was dealing with. Her work is interesting in intent, but fundamentally deficient and wrong in execution.

The second difference concerns her basic hypothesis. At the beginning of her work, in the first memoir, Sophie Germain had a very vague notion of what elastic force or elastic moment might mean in Lagrangian mechanics; hence her demonstration was in error not simply because of its mathematical deficiency, but also because of conceptual ambiguity in the formulation of her hypothesis. She now abandoned her previous attempt to justify the hypothesis that elastic force is proportional to the change in curvature. She had tried this through consideration of the geometry of deformation of four points on the surface, an argument that Legendre in 1813 had declared unsatisfactory. Now she simply argued *ab initio* that this elastic force was proportional to the *difference in shape of deformed and undeformed surfaces* and thus, since shape is determined by curvature, proportional to difference in curvature.

To establish an expression for the curvature of a surface was not a simple problem. In the case of the beam, which Sophie Germain cited in support of her argument, there is little difficulty; the deformed beam is represented by one curve in space, with one radius of curvature. The elastic deformed surface, however, presents a multitude of possible curves through any point on the surface.

There was a way out of this difficulty that Sophie Germain knew and used. If the curvature associated with two particular perpendicular planes are known, all other curvatures can be obtained from them. She used this property of curvature of surfaces to define the shape of deformed and undeformed surfaces, thereby placing elastic force proportional to:

$$\left(\frac{1}{f} + \frac{i}{g}\right) - \left(\frac{1}{F} + \frac{1}{G}\right)$$

where F and G represent the two principal curvatures of the undeformed surface and f and g those of the deformed surface.

It is probable that Sophie Germain was moved to broaden the scope of her analysis by Poisson's 1814 memoir, or at least what she could glean of it from its publication in abridged form.[3] Indeed, she not only set herself the task of establishing a theory for initially curved surfaces (based on her own hypothesis), but also she strove to extract from her analysis the nonlinear equation that Poisson had presented in his memoir. This she did in the course of her exercise. Here roles were reversed: Sophie Germain derived an equation of Poisson's by her own method, as he had derived her equation. But while he was drawn to the equation in her memoir by the analytic and experimental evidence that suggested its correctness, she was drawn to Poisson's equation by his professional prestige. However understandable this may be, his nonlinear equation was wrong, hence hers was wrong, and her derivation was erroneous.

The final section of her memoir described the experiments she had been making with curved surfaces. She had been trying to excite and make visible the mode patterns of cylindrical surfaces the way Chladni had done using flat plates. His technique, however, when applied to initially curved surfaces, is no longer so simple, and Sophie Germain, not surprisingly, met with enormous difficulties. What she was trying to do, of course, was to demonstrate experimentally that her generalization of the plate equation was accurate, that fact conformed to theory as well as theory conforming to fact. The report she gave of her experiments is more narrative than expository, honestly describing the limitations of her accomplishments as well as their partial successes. Her tone is appropriately tentative, and she concluded:

C'est avec regret que je me vois forcé de terminer ici mes recherches. Je sens combien elles laissent encore à désirer. Mais peut-être aussi m'eut il fallu plusieurs années pour rassembler, relativement aux seules surfaces cylindriques, une masse de faits qui permis d'agir sur ce genre de surfaces avec le dégré de certitude que l'on a obtenu dans le cas des surfaces planes, à l'aide des nombruses expériences de Mr. Chladni.[4]

All in all, this third memoir was much thinner than the second. She did not repeat what she had done before, though she did refer back to it.

From a modern point of view, more than a contemporary one, her manner of justification of her hypothesis represents an interesting departure from her previous work. The variational approach applied to a function of the principal curvatures, a function representing the 'elastic moment' in Lagrange's sense has a curiously modern ring to it when applied to a plate-type structure. It is an approach that leaps over all microanalyses: internal forces, including stress, are entirely neglected; whatever may be going on inside the surface, elastic behavior is measured by deformation. A modern method of deriving this equation with boundary conditions requires the notion of invariants of a curved surface as well as an understanding of the variational technique, but it also considers internal forces as secondary. Sophie Germain's work cannot be said to anticipate this modern approach, since the rest of the nineteenth-century work that led to present elasticity theory followed a far different course. Yet her assumptions, her way of looking at the problem, seem familiar to the modern eye, a familiarity lacking in the contemporary molecular approach of Poisson.

This part of Sophie Germain's memoir was not going to interest the commission of judges, but her experimental efforts provided grounds for awarding the prize. The commission, consisting of Poisson, Laplace, Legendre, Poinsot, and Biot, included this statement with the award:

The class has received only a single memoir, a sequel to that which obtained an honorable mention in 1814 and to which the author has added new developments. The differential equation given by the author is correct although it has not resulted from the demonstration. Yet the manner in which the particular integrals satisfying it have been discussed, the comparison made with the results observed by M. Chladni and finally the new experiments attempted on plane and curved surfaces in order to test the indications of the analysis, appear to merit the award of the prize. . . . The author is Mademoiselle Sophie Germain of Paris.[5]

The reservation contained in this statement provoked this letter to Poisson from Sophie Germain:

[January, 1816]

The judgment pronounced by the class taught me that I had been deceived by the proof that had been acceptable to you, but it did not explain to me the nature of the error I made. M. Halle, to whom I expressed my curiosity about the mistake in my proof was willing to ask you to clarify my doubts. I do not think I was mistaken in the manner in which the general equation was deduced from the hypothesis; therefore it would have to be the hypothesis itself that was not justified in a satisfactory manner.

In order to spare you the bother of reviewing the proof, I have reproduced in the adjoining note the arguments on which it is found. I have indented them so that it will be easier for you to mark where you think the chain of reasoning falls apart.

The more respect I have for your judgment, the more importance I attach to obtaining the clarification I solicit.

Do accept, Monsieur, the assurance of my most distinguished consideration.

> Whatever the nature of the forces considered, they are proportional to the effect they produce or tend to produce.
>
> The forces of elasticity tend to destroy the differences between the natural shape of the bodies endowed with this force and the shape that those same bodies are forced to take by an external cause.
>
> The forces of elasticity acting in any elastic body are therefore measured by the difference in the natural shape of the body and the shape that an external force would cause it to take.
>
> The effect produced by a force is implicitly or explicitly the sum of the effects produced by the same force: explicitly, if one successively considers all the diverse effects taking into account their inter-dependence; implicitly, if the connection existing between these same effects permits them to be considered as a single thing.
>
> The effect of the forces of elasticity acting in a surface is to destroy, [minimize], the difference between the natural curvature of the surface and the curvature that the same surface is forced to take through the action of an exterior cause. But the question of curvature of a surface cannot be answered simply: it is composed of the group of questions relative to the curvature of curves resulting from sectioning the same surface in all directions and under every possible inclination.
>
> The sum of the differences between the curvatures of the curves formed by the various sections of the surface, considered before and after the action of the exterior force, is therefore explicitly the measure of the forces of elasticity acting on this surface.
>
> There exists between the curvatures of the curves formed by the various sections of the surface a relationship such that it is permissible to express their sum by that of the principal sections only.
>
> The effect of the forces of elasticity is then implicitly expressed by the sum of the differences between the principal curvatures of the surface alone, considered before and after the action of the external cause.[6]

The judges had in fact criticized Sophie Germain for the way she had moved from her hypothesis to its justification, i.e., for her mathematics. She, however, took their criticism as a judgment of her hypothesis and used her macroanalysis as a stance from which to address Poisson, asking him to argue his hypothesis against hers. Poisson, of course, was not interested in doing any such thing. He replied to Sophie Germain by refusing her request for a critique, stating the reason for the award with reservation, and offering to send her a copy of his memoir when it appeared in print:

Mademoiselle,   Paris, 15 January, 1816

M. Halle has just delivered a letter that you do me the honor of addressing to me, which contains several questions relative to your memoir. The reproach that the commission made concerns not so much the hypothesis as the manner in which you applied the calculus to the hypothesis. The result to which these calculations have led you do not agree with mine except in the single case wherein the surface extends itself infinitely little from a plane, be it in a state of equilibrium or of movement. My memoir will be printed shortly and I contemplate offering you a copy, as soon as the printing is finished.

Permit then, Mademoiselle, that we adjourn this discussion until the time you will have been able to compare my results with yours.

Accept my respects and my high consideration.

Poisson[7]

CHAPTER EIGHT

PUBLICATION

In 1816, Sophie Germain found herself in a new position. She had spent five years in almost single-minded concentration on the plate problem and had been awarded the *prix extraordinaire*. On the one hand, this gave her a sense of professional standing, authority, and self-esteem. She had been the one – for a while the only one – doing fruitful work in the investigation of the elastic behavior of surfaces; now her work had won a measure of public recognition.

On the other hand, the core of the scientific community, Poisson particularly, treated her without the respect she felt she deserved. He did not deign to engage in professional discussions with her; in addition he claimed precedence in plate theory. Years earlier she had been willing to see herself as the lowliest novice in the company of the great, but now she had faith in her own competence and no great admiration for the contribution of her rival.

Two events of 1816 reinforced her self-esteem. One was her meeting with Joseph Fourier. His friendship and position in the Royal Academy of Sciences gave her the sense of real participation in the activities of the Parisian scientific community. More important for her actual accomplishment, however, was her return to serious work in number theory, a field of inquiry she had neglected during her engagement with vibrating surfaces.

The same session of the First Class that had named Sophie Germain as the recipient of the *prix extraordinaire* established a new contest in mathematics.[1] Its object was Fermat's Last Theorem: there are no integers x, y, and z which satisfy the equation

$$x^n + y^n = z^n$$

where n is any integer greater than 2. Sophie Germain eagerly accepted the challenge posed originally by Fermat. She certainly did not prove the theorem in its complete generality; indeed, it still remains an enigmatic problem. In 1816, however, the expectation that a proof might be formulated did not seem outlandish. Fermat had sketched a proof for the case $n = 4$. Euler had shown the impossibility for the case $n = 3$. The case $n = 5$ was soon to be solved by Legendre.

Sophie Germain's work resulted in a contribution rightly acknowledged by mathematicians. What she accomplished is embodied in a theorem (now carrying her name) with which she showed that for all prime numbers n less than 100, there are no solutions to the above equation for the case in which none of the three numbers x, y, and z is divisible by n. At a time when so little had been achieved with respect to Fermat's Last Theorem, her contribution was clearly significant.[2]

The subject of this prize, after being reset in 1818, was retired in 1820, since there were still no entries worthy of the award.[3] There is no evidence that Sophie Germain submitted any of her work to the Academy for judgment over this four-year period. She had evidently had enough of competitions. However, a letter to Gauss dated May, 1819, indicates that she was earnestly engaged in exploring the impossibility posed by Fermat:

Although I have labored for some time on the theory of vibrating surfaces (to which I would have much to add if I had the satisfaction of making some experiments on cylindrical surfaces I have in mind), I have never ceased to think of the theory of numbers. I will give you an idea of my preoccupation for this kind of research in avowing to you that, even without any hope of success, I prefer it to work that necessarily gives me results.

A long time before our Academy proposed as the subject of a prize the proof of the impossibility of Fermat's equation, this challenge, transmitted to modern theories by a mathematician who was deprived of the resources we possess today, has often tormented me. I *vaguely* perceive a liaison between the theory of residues and this famous equation. I believe I have spoken to you in the past of this notion, since it struck me as soon as I read your book.

Here is what I have found:. . .[4]

Sophie Germain then proceeded to describe her efforts with respect to the subject of the current prize in mathematics. The details of her letter show that she had not, in 1819, given final shape to her work. A few more years and Legendre's assistance were required for this. Assorted drafts of her work, as well as her letter to Gauss, also reveal that she initially sought to produce a general proof of the impossibility of Fermat's equation.[5] Legendre acted to constrain her desire for the broadest possible applicability of her approach. Working together, they molded her ambitious ideas into a more limited, but rigorous proof of substantial value. This is the only instance in Sophie Germain's career where she had an opportunity to work closely as a professional equal, with a colleague in a productive way. (Legendre's role in the plate problem was clearly more that of a patron than peer.) The sole report of her achievement is found in

Legendre's memoir, "Recherches sur quelques objets d'analyse indéterminée et particulièrement sur le théorème de Fermat," where it appears in a footnote.[6] From a twentieth-century mathematical perspective, this was the most important work she did, even though the publicity accorded her name as recipient of a *prix extraordinaire* has been largely responsible for the preservation of her memory.

Legendre at this time was in the twilight of his career. Though he could welcome conversing with Sophie Germain on matters of number theory, he would be eager to avoid further involvement with plate theory. A new arrival to the Parisian scientific establishment, Joseph Fourier, took over for him in this respect. Fourier had returned to Paris in 1815 at the nadir of his fortunes.[7] When and how he met Sophie Germain is unknown. The first evidence of their acquaintance is a letter from 1816:

Mademoiselle,   Thursday 2 May [1816]
 I have the honor of proposing to your mother and you a luncheon with me next Sunday, along with Madame de Bressieux and Madame de Vauborel. I pray you will excuse the foolishness of my project and to consider only its intention, the excuse of all mediocre authors. I certainly do not have the means of receiving you with as much worthiness as I would like, and I feel that I am asking a sacrifice of you, but I would be very grateful if you would have the goodness to present my invitation to your mother, asking her in your name and mine to accept.
 Monsieur and Madame de Prony will be with us. There will be no others, except my secretary and me and perhaps the nephew of Madame Vauborel, if he is in Paris. We will meet at a quarter after twelve.
 I send you also an engraving of which I have several copies. It is the portrait of one of our predecessors, one of the most illustrious founders of a science that you love and have enriched.
 Agree to accept, Mademoiselle, with my respect, the invitation which I make, and grant me this kindness as a new witness of the gracious obligation to which you have made me accustomed.

                                                                    Jh. Fourier[8]

This letter came near the beginning of a friendship that lasted until at least 1827, the year of his last preserved letter.[9] What brought them together seems to have been more personal than professional. Unlike most of their colleagues, both were unmarried. This provided a bond, not from romantic possibilities (of which there is no suggestion), but from their mutual devotion to the single state and their comfort within it. Both of them also seem to have been kindly, rather peaceable people. Fourier, whose life had involved so many public and political tensions and embarrassments, might well enjoy the society of an intelligent, thoughtful, and secluded person, for whom all the turmoil of his past was somehow not very real and perhaps not very interesting.

They shared, also, immediate experiences. Arago reports in his eulogy that Fourier too, in his youthful enthusiasm for mathematics, while a student at a military school directed by the Benedictines, foraged in the kitchen and corridors for the remains of candles to shed light on his solitary nocturnal studies.[10] Then, too, Fourier's professional respectability, like that of Sophie Germain had been affected by Poisson. In 1807 Fourier had submitted to the Institute a "Mémoire sur la propagation de la chaleur," a treatise that now is regarded as one of the most significant productions of applied mathematical thought of the Napoleonic period. The commission appointed to review Fourier's work, consisting of Lagrange, Laplace, Monge, and Lacroix, questioned the solidity of his analysis in several respects. Poisson wrote the published statement of their reservations (even though Poisson was not yet a member of the First Class).[11] Fourier also had ambitions with which Sophie Germain could help him, and he, in turn, could be of help to her. In 1822, the year of Fourier's election to the position of Permanent Secretary of the Royal Academy of Sciences, we find this letter:

Mademoiselle,   Friday morning [prior to Nov.] 1822

I cannot adequately express how much I acknowledge the interest you accord me and the perfect grace with which you express it. Those you love and favor cannot be unhappy. Allow me to suggest that you not go out; the air is cold and a great number of people are indisposed. I returned on foot from the Faubourg St. Honore on Tuesday evening; I was seized with a cold that caused vivid discomfort in my entire body. The physician, understanding the language of the geometer, calls this indisposition a curvature. Mine must certainly have been of a very high degree. Finally it ceased entirely and I was able to go out.

I cannot doubt now that the wish of the largest number of my colleagues is to choose me, and the rival of mine who flatters himself so much is in great error. But he has so many artifices to resort to that it would be unwise not to be concerned with him.

M. Desfontaines has told me that M. Legendre has discussed this election with him and that, without denying the interest he took in M. [Biot], he assured him that he would give me his support in every possible way. M. Desfontaines appears convinced. I do not doubt, Mademoiselle, that your eloquence touched him. Any support that I owe to you remains more valuable in my eyes. That of M. de Jussieu is very honorable in itself, and I do not doubt that you inspired it.

The election will not take place until the beginning of November. I am surprised that at this time M. de Jussieu will still be in the country. I am sorry to learn of the absence of M. de Montmorency, because his opinion would have been favorable to me. In the end the gods will decide. But my sentiments of acknowledgement and respect are independent of the gods.

I beg you to accept them.

Fourier[12]

Fourier's rival in this contest was Biot. His victory over this member of the Society of Arcueil and admirer of Poisson's accomplishments stands, from the perspective of history, as a political statement reflecting the declining status of Laplacian physics – a far-reaching decline that affected all scientific endeavor, including elasticity theory. At the time, the personal aspects of the election were far more obvious. For Sophie Germain it meant a well-earned reward.

Institute of France
Royal Academy of Science
Paris, 30 May 1823

The Permanent Secretary of the Academy
Mademoiselle,

I have the honor of informing you that every time you wish to attend the public meetings of the Institute you will be admitted to one of the reserved seats in the center of the hall. The Academy of Sciences wishes to demonstrate, by this distinction, all the interest that your mathematical works inspire, especially the scientific research that it has crowned through the award to you of one of its annual, grand prizes.

Accept, Mlle., the offer of my respect.

Fourier[13]

Thus as one of his first official acts as Permanent Secretary, Fourier made it possible for Sophie Germain to attend all public sessions of the four Academies comprising the Institute.[14] Two days later, he wrote an informal letter, complementing his official communication, inviting her to his first public session as Permanent Secretary, and revealing his nervousness at having to speak before this formal assembly of colleagues, their guests, and assorted dignitaries.

Sunday, 1 June 1823

I have the honor of recalling myself to the memory and good will of Mademoiselle Germain. I have wished for some time to present myself to her, but urgent business has kept me from this. I send her along with this: (1) an official letter; (2) a center ticket for the person accompanying her. If Mlle. Germain does not plan to attend the meeting, I ask her to dispose of the ticket in the way she sees fit, and if it should be necessary I could deliver one or several others, but not for the center.

Alas, I would rather keep all these tickets. I am condemned to cause the public a great annoyance, and I will appear tomorrow like a faint gleam in the middle of a display of fireworks. But I am resigned to all possible comparisons. From the start it has seemed reasonable for me to take as grave and simple a tone as I can maintain, and to abstain from every pretension to success that I will not be able to attain and barely desire. That which I desire above all is to keep the esteem and the remembrance of Mademoiselle Germain.

I ask her to receive the expression of my respects.

Fourier[15]

The value of Fourier's promise of access to the annual public sessions of the Academy of the Institute can be surmised from the following letter, written to Sophie Germain two years earlier by Delambre, Fourier's predecessor as Permanent Secretary. It demonstrates not only her desire to attend these sessions – a desire that could safely be assumed – but also the difficulty she had gaining entrance to the public meetings, prior to Fourier's election.

Mademoiselle,   Paris 25 July, 1821
    . . . For several years now, the established order for the distribution of places at public sessions seems to me, as well as to you, susceptible to more than one objection. The cause of this situation is the late M. Suard, and there is not a member of the Institute, especially if he is married, who has not complained of it at one time or another. According to a royal ordinance, however, each Academy can dispose of its places according to its own conventions. L'Academie Française having seized all the choice places on the day when it presides, each one of the other academies wished to be privileged in turn. The use of the *billets du gentre* has been retained. They are distributed amongst the President, the Vice President, the two permanent secretaries, and the readers. A number of these tickets are reserved for grand functionaries and celebrated foreigners, so that at each meeting of our Academy I have at most ten such tickets. I make it a rule to distribute them to those of my fellow members who want them for their wives. *Le Bureau* has no more than 40 tickets to distribute and the academies have 75. Happily their wives are not so eager to attend, so that up until now I have not had to refuse a crowd. It is to the perpetual members and not to those who change each year that foreigners prefer to address themselves. But if I had been informed of your desire to attend our public sessions, I would have found the means to reserve for you one of these tickets, and I will certainly do so next March. It may not be impossible in July; the permanent secretary of the Academy of Inscriptions usually grants me several tickets for the center. For the two other academies, however, I do not have a single one, and if my wife had not renounced these solemnities a long time ago, I would be obliged to approach the functioning permanent secretary on her behalf. I hazarded it only once and my request was accepted; I like to think that it would be every time, and at present I assume this to be the case. I would voluntarily run the risk every time you would like one of these tickets, if you inform me of your desires, at the latest, on the Tuesday before each meeting.

    Accept the offer of the respectful sentiments with which I have the honor to be, Mademoiselle, your humble and very obedient servant.

Delambre[16]

All of this fussing over tickets and wives! How outrageous it must have seemed to Sophie Germain, who now felt herself a professional mathematician with a legitimate claim to a place among her peers. How doubly frustrating to have to cope with this kind of exclusion while she aspired to confront Poisson directly in some way, her hypothesis against his hypothesis, to lay absolute claim to real accomplishment.

It was during this time that Sophie Germain was preparing her prize-winning work for publication. Lacking any formal endorsement on the part of the Academy of Sciences, she published it privately at her own expense in July 1821.[17] Legendre had suggested in 1814, upon her receipt of an honorable mention, that she publish her work. Now he encouraged Fourier to help her with it. A letter indicates that Fourier did consult with Sophie Germain on the substance of her research before its publication:

Mademoiselle,   Thursday morning, 1 June [1820]
Monsieur Legendre has wished on your behalf that I undertake a study of a memoir on the properties of elastic surfaces. I have very attentively read this work and I have found there new proofs of the success of your research on this difficult problem. I propose that I have the honor of coming to your home the day after tomorrow, Saturday, at eight thirty in the evening and of giving you an account of my thoughts on the subject. This hour has been indicated to me as the most convenient for you. If you prefer another hour, or another day, I pray you will have the goodness to inform the carrier of this letter. I will have the honor of presenting myself Saturday.
Agree, Mademoiselle, to the homage and respect of your very humble and obedient servant.

Fourier[18]

Two other letters written near the time that her manuscript was assuming final form suggest that he was far from eager to become deeply involved with her work:

Mademoiselle,   Paris, 19 September, 1820
I extremely regret leaving for the country without having had the honor of seeing you. I will remain there about fifteen days, and on my return I will visit you at your home, if you would permit me. I am convinced that you will have developed further the theory with which you occupy yourself, and which I invite you to publish right away. No one can treat this difficult and interesting problem with more success.
Accept, Mademoiselle, the offer of my respect.

J. Fourier[19]

Mademoiselle,   [November, 1820]
I acknowledge your remembrance and your goodness. I intended every day, since my return from the country, to have the honor of seeing you. Each time I was prevented by some new incident.

I certainly accept the invitation of Madame your mother for Wednesday, December 6, and I hope that she permits me to offer her my respects in the course of that week. I long to take up again our analytical conversations, because you bring to them much intelligence and a marvelous wisdom. I beg your pardon for this remark because praises displease you. I promise to attempt to correct myself.

I pray you, Mademoiselle, to offer to your mother and to accept yourself my respects.

Jh. Fourier

There was good reason for Fourier's maintaining his distance. In her quest for the prize and professional recognition, Sophie Germain had broadened the scope of her analysis to include the behavior of curved surfaces. She had, in choosing to address this more difficult problem, moved into truly uncharted waters. Equipped only with a tenuous grasp on relevant physical concepts, and a less than full understanding of the substance and technique of the variational approach of Lagrange, yet confident in her abilities, she produced a memoir that is a complex mixture of maladroit adaptation of accepted principles, arbitrary (yet seemingly reasonable) assumptions, sometimes correct, sometimes faulty use of variational technique, and a lengthy discussion of the implications of her analysis for the vibration modes and frequencies of circular curved beams and cylindrical surfaces. To evaluate her endeavors, to ferret out the true and rectify the false, would have required Fourier to set aside all other activities for a goodly period of time.

Yet Fourier did devote some attention to one familiar element of her work. Sophie Germain, as reflected in her letter to Poisson after receiving the award, had understood the judges to mean that the difficulty in her analysis lay in the demonstration of her hypothesis itself, rather than the derivation of the plate equation proceeding from that hypothesis. Her problems concerning the derivation were so intimately tied to her irremediable limitations that it is no wonder Fourier did not want to become deeply involved. He evidently did try to help her, however, with a purely geometrical demonstration justifying her basic elastic force-principal curvature relationship. She acknowledged his contribution in her preface in the following way:

The Class accorded the prize to my memoir, but announced, that my demonstration had not appeared entirely satisfactory.

Since that time, I have occupied myself, on various occasions, with the theory of elastic surfaces. I have multiplied my experiments, calculations, and reflections. I confess that I have continually thought of new reasons to hold to my opinion.

I was preparing to publish my thoughts when M. Fourier desired very much to come to understand my demonstration. This intelligent judge stated that he preferred a purely geometric demonstration to the reasoning upon which I had relied. He proposed to me, as a model, the way that Jacques Bernoulli had, at another time, proceeded using the hypothesis that applied in the case of a straight beam.

This demonstration seemed to have fallen into oblivion; I had totally ignored its existence. Based on the way in which Euler and, above all, Lagrange had expressed it (*Mécanique Analytique, Mém. sur les ressorts, ployes*), I had always believed that this hypothesis had only been introduced as a purely arbitrary assumption. Bernoulli's memoir appeared obscure to me. M. Fourier extended his kindness as far as taking the trouble of explaining it to me. It pleases me to point out that, without the help of this skilful geometer, I would never have appreciated, nor even understood, the demonstration of the author.[21]

Fourier had tried to help her understand Jacques Bernoulli's demonstration of the moment-curvature relationship appropriate to the beam flexure problem, but her adaptation of Bernoulli's demonstration to the plate flexure problem appears as naive and unconvincing.[22] Her difficulty here, as with variational methods, seems to originate in her lack of rigorous training from the beginning. Now she had come too far on what she had taught herself to be in a frame of mind to study the basics as a beginner, even had anyone been willing to offer such instruction. She did not clearly recognize her deficiencies, and hence would be a difficult student in any case. It was easier for a friend like Fourier to put her off than to grapple with these problems.

Thus she published her work without being aware that it displayed her incompetence as well as gave evidence of professional respectability. It was for the sake of this respectability above all that she had to publish. Without this public statement, future generations might only be able to cite and study Poisson's 1814 memoir as the fundamental basic contribution to the development of a theory of the elastic behavior of plates. Sophie Germain's sensitivity to this issue appears in the following paragraph, taken again, from the preface to her treatise:

The theory that I seek to establish is not yet known to the public. The only demonstration of the equation of elastic plates that has been published to date has been given by the author of the "Mémoire sur les surfaces élastiques," and I am not able to conceal the fact that the principles upon which it is based are not absolutely irreconcilable with those that I have been led to adopt. If it were a matter involving an obscure author, I would limit myself to presenting the problem as I conceive it. This is far from the case; the geometer with whom I have the misfortune of not sharing an opinion, has such a right to respect, that the authority attached to his name yet stays my own judgment. I believe that I would have hidden from the reader the strongest objection that one could make against my hypothesis if I did not admit that it differs in its entirety from that due to this learned author.[23]

Though Poisson was in the position to ignore her publicly, she could turn her obscure status to her own advantage, using it as the base from which to make an appeal to a wider public, a broader world of science that should judge between their claims. Her preface initiated this appeal:

Acoustic phenomena, the knowledge of which we owe to M. Chladni, has directed the attention of geometers to the problem of elastic surfaces.

Despite their work, there remains some embarrassment over the choice of principles that ought to serve as a basis for this theory.

Two essentially different hypotheses have been proposed. One of them has the support of a justly celebrated name; that is a strong reason to distrust the other, which belongs to me. Indeed, I too have made a total effort to renounce it.

I feel that I have the need of support of an independent judgment. This is the only means remaining to me to dissipate the doubt that has followed me throughout the research that I have undertaken. In exposing my feelings, I add my reasons in order that an informed public may weigh them and advise me of a better judgment.

If sometimes I speak in an affirmative tone, it is only to free myself from the tiring expression of doubt. It suffices to warn the reader this once that, far from intending to fix his opinion, I am asking, on his part, for a critical examination of mine. You will excuse me, without doubt, for not hiding then from anyone the advantages that I believe to recognize in my hypothesis.[24]

Sophie Germain saw herself involved in an intellectual struggle with Poisson, a struggle that could engage the interest of others in the scientific community, and probably provides the reason she was encouraged to publish work done five years previously. She became, here, part of the historical trend that was challenging the molecular mentality and moving away from it. Indeed, she did address one true limitation of the corpuscular scheme. Working with sensible forces at insensible distances renders the task of formulating boundary conditions for the plate well-nigh impossible. Poisson had promised to do this, but had failed to fulfil his promise. Sophie Germain, on the other hand, could claim that the boundary conditions, as well as the differential equation defining the motion of interior points of the plates, flow from her hypothesis and the variational approach. Here she was on solid ground, at least in principle. An extended excerpt gives witness to her critical facility in the matter:

I have awaited the publication of the determination [of boundary conditions] on the part of the author for a long while. In the interest of the problem, I would have desired that he himself develop all the consequences of the hypothesis that he has adopted. . .

If a hypothesis contains all that is part of the problem, if it can be regarded as a true definition, it suffices to introduce this hypothesis into the calculus, in order to obtain all the analytical consequences that belong to the solution of the same problem. The determination of the state of the points at the boundary of the surface is no more foreign to the problem than the determination of the state of the interior points of the same surface; integration by parts, then, ought to give, at one and the same time, the terms that apply to the boundaries and those that enter into the equation of the surface. . . .

The hypothesis that I have proposed fulfills this condition; and, in those cases that have permitted me to establish a comparison of theory with experiments, the state of the points at the boundary has not been less justified than that of the interior points of the surface. As a result, my hypothesis can be regarded as equivalent to a true definition of elastic force.[25]

On this general level of criticism Sophie Germain could claim a degree of respect that was unmerited by the details of her mathematical analysis. This distinction was never made clear to her, however, nor was she given a clear understanding of her limitations. Instead she was encouraged to think well of herself and her work. In retrospect it becomes clear that general flattery was taking the place of the serious critique with which she had occasionally been honored in the past. The reception accorded her published memoir is indicative: it drew expressions of praise from Cauchy and Navier, both respected mathematicians, from Delambre, writing in his official capacity as Perpetual Secretary of the Academy, and from her aging friend, Legendre.

Cauchy's note is aloof, quite formal, and does not acknowledge her accomplishment, only her name and topic:

Mademoiselle,   Paris 24 July, 1821

I have received the work that you have had the goodness to send me; the name of its author and the importance of the subject equally recommend it to the attention of geometers. I am, at the moment, only able to offer you in return a volume in which I have sought to clarify the principal difficulties of algebraic analysis. Please accept, I pray you, the homage of my distinguished consideration and very humble respects.

Cauchy[26]

Navier indicated that he had actually read her work with considerable interest:

Mademoiselle,   Paris 2 August, 1821

I have gratefully received the work which you deigned to send me. Reading it has inspired me with great interest, and I appreciate, equally, the value of a composition so remarkable, one that few men can read and that only a woman could write.

I have the honor of being with respect, Mademoiselle, your very humble and obedient servant.

Navier[27]

Navier's praise of Sophie Germain's style is justified; her preface is an unusually open and honest expression of her experiences in plate theory and prize winning. "A composition . . . that few men can read and that only a woman could write" might possibly be read as a snide comment, but surely that would be anachronistic and most unlikely for Navier in any case.

Delambre indicated that her memoir would be honored by its inclusion in the library of the Institute, a questionable honor, if any at all.

Paris, 23 July, 1821
Perpetual Secretary of the Academy to Mademoiselle Sophie Germain
Mademoiselle,

The Academy has received with the greatest interest the work that you have seen fit to address to us, which is entitled: "Récherches sur la théorie des surfaces élastiques" and which you have published. The Academy in expressing its recognition of this new proof of your talents, has instructed me to thank you in its name for having submitted this interesting memoir, which it has had the honor of placing in the Library of the Institute.

Agree, I pray you, Mademoiselle, to the homage of my respect.

Delambre[28]

Legendre accorded her a more substantial greeting, but excused himself from any commentary concerning the technical content of her work. He had always remained at a distance from elasticity.

Mademoiselle,   Paris, 23 July, 1821

Last Tuesday I received the memoir you chose to send me, a memoir on beautiful paper, beautiful cover, and a short, most obliging, but too modest, letter. I give you my very sincere compliments for having finally triumphed over your repugnance toward making public these investigations that have cost you so much effort. I hope that you do not find occasion to repent of your courage, and that this first publication, which was the most difficult, will soon be followed by several others that, no doubt, will obtain for you the esteem and support of *connoisseurs*.

I have been able to peruse only the first pages of your memoir, and you may well suppose that I am far from being able to render judgment on this work, which is one of those that can only be appreciated through lengthy and profound study; undoubtedly, you yourself are repelled by any judgment founded only on a superficial examination.

I have found your preface very well written. It very nicely reviews the status of the problem, you advance your opinion in a most modest manner and, if there is anything to reproach you for, it would be the compliments with which, in some ways, you praise the geometer with whom you contest your opinion. That he is able to respond with dignity to this civilized assault I desire more than I hope.

I was surprised not to see in the *Errata* the word *campanarum* in place of *campanorum*, which, unfortunately, is repeated three or four times.

As soon as we are able to take a short stay in Paris, we would be eager to have the honor of

seeing you. My wife is fine now. She sends a thousand tender compliments to which if you
permit me, Mademoiselle, I add my respective homages.
Legendre[29]

It is hard to imagine that Sophie Germain could receive these letters as
anything but deserved praise, even if their vacuity left her somewhat
uncomfortable. It is impossible to imagine a man in the profession being
treated this way. With no way of avoiding it, and through no fault of her
own, she looks like a fool in the world of French science. This is especially
poignant in the context of the new work that was appearing in elasticity
theory.

CHAPTER NINE

EMERGENCE OF A THEORY

The first new name associated with elasticity was that of Fourier. An 1818 note shows how a solution to the plate equation could be obtained through an approach uniquely associated with his name: through the representation of plate motion in the form of a summation over the trigonometric functions, sine and cosine. His focus was on solution technique, not on a demonstration or generalization of the plate equation. The particular problem he solved also avoided a deficiency in theory that existed at that time, namely, the lack of an accepted set of boundary conditions. Since he intended to describe the response of an *infinite* plate to a quite arbitrary initial disturbance, the need for knowing conditions at an edge did not arise.[1]

This note represented the extent of Fourier's interest. It complemented rather than competed with the major thrust of Sophie Germain's work, and really added nothing to knowledge in elasticity. It simply used the plate equation for an application of a mathematical technique. Fourier's colleague, however, a former student and a contemporary of Sophie Germain, C. L. M. H. Navier turned serious attention to the "physics" of the plate problem. His first memoir, read to the Academy at its session on the fourteenth of August, 1820, had limited distribution, as it was available only in a small privately lithographed edition.[2] Nevertheless, this "Mémoire sur la flexion des plans élastiques" stimulated an interest in elasticity that encompassed physical phenomena that fall within what is now termed the general theory.

Navier turned to the theory of the elastic plate with the motivation of one schooled in the practical arts, i.e. as an *ingénieur*. His concern was with the behavior of a floor slab, supported at its edges and loaded in a rather arbitrary manner. His solutions defined the deflections of any point of the plate, the reactive forces at the plate's edge supports, and finally, the load at which the plate may rupture. It is a problem of "static equilibrium" rather than of motion, of dynamics. Navier's introductory paragraph described his subject and, in addition, gave the reader a brief and accurate summary of the contributions of various people to the development of a plate theory.

The equation of equilibrium for an elastic plate, a fourth-order partial differential equation, was presented by Lagrange without proof. M. Poisson derived this equation through the consideration of molecular actions that act over very small distances. A person, whose work has been crowned by the Academy of Science, and who cultivates, with distinction, those sciences which ordinarily remain foreign to her sex, deduced from this equation an explanation of several phenomena observed in the vibration of plates. M. Fourier has discovered by his method – in giving for the first time the complete integral of this fourth-order equation in a form sufficiently simple so that one can see in the mathematics what is physically happening – the nature of the motion of an elastic plate of infinite extent due to any small initial change in the shape of the plate. We propose to deduce from this same equation the complete solution for the [static] equilibrium of an elastic plate. This problem is of interest in analysis since it supposes the knowledge of new integrals of a fourth-order partial differential equation. It is also of interest in the arts of construction, because its solution gives laws for the design of diverse kinds of floors.[3]

Navier's interest in the behavior of a plate of finite extent forced him to reconsider the derivation of the equation Lagrange had provided without demonstration. He had to establish appropriate boundary conditions. The approach he took was that embodied in the *Analytique Mécanique*. But whereas Sophie Germain's mastery of the variational technique was clearly deficient, Navier displayed full competence. In addition, he provided a rational and clear (yet conceptually deficient) derivation of an expression for "elastic moment"; a prerequisite to the application of the variational technique.[4] His process yielded a consistent, yet not fully correct, set of boundary conditions as well as the equation given by Lagrange. With these in hand, Navier proceeded to apply Fourier's series technique to obtain solutions to several practical problems.[5]

Navier read his memoir before the Academy at its session of 14 August, 1820, and Prony, Poisson, and Fourier were appointed to evaluate his work. This commission's composition was not entirely to Navier's advantage since, although Prony and Fourier would have been sympathetic in their judgment, Poisson, confident that he had, in 1814, established the appropriate framework for the analysis of plate behavior, would have viewed Navier's work as a challenge to that accomplishment. Navier's analysis is free of any molecular notions, i.e., of sensible forces at insensible distances.

Poisson's presence on this commission was not the only impediment to a swift and favorable review. Eventually, Cauchy, whose work is rightfully regarded as of prime importance in the development of the theory of elasticity, was added to this commission and designated as *rapporteur*. Cauchy, to the dismay of many, often became intrigued with work he was expected to report on and delayed official pronouncement of the Academy while he pursued his own investigations in the same field. This he did also with Navier's memoir.

With Poisson and Cauchy responsible in part for the evaluation of his endeavors, Navier was left waiting for approbation. He did not remain idle, however, for within a period of eight months he fashioned and sent on to the Academy another memoir for consideration. Again, Prony, Poisson, and Fourier were selected at the Academy's session of 14 May, 1821, to judge this new piece of work bearing the rather cumbersome title "Mémoire sur les équations différentielles qui contiennent les lois des déplacements des molécules des corps solides élastiques lorsque ces corps sont maintenus en équilibre sous l'action de diverses forces, ou vibrent par suite de l'action de ces mêmes forces." In the view of some historians, this memoir marks the birth of today's theory of elasticity.[6] It contains the first formulation of a system of differential equations defining the way the points in an arbitrary, three-dimensional, elastic solid move when acted upon by external forces. Previous analyses of the elastic behavior of different structural shapes, e.g., plates and beams, were limited in their applicability. Navier had developed a theory that, in principle, was applicable to the explanation of all elastic phenomena. Poisson must have realized the potential significance of Navier's accomplishment; however, there was more than this to disturb his pride, for on this occasion, Navier adopted the molecular conceptual scheme as a basis for analysis. Poisson had been upstaged in his own domain. Furthermore, in contrast to Poisson's laborious style, Navier's work is short, lucid, and displays a heightened awareness of the intimate relationship between physical behavior and mathematical approximation.

Navier, from the outset, restricted his attention to the case in which an elastic medium experiences small deformations. If Poisson had done the same in 1814 he would not have produced, or rather reproduced, the plate equation. Navier also assumed that the force generated between any two molecules is proportional to the change in distance between them. Poisson would later declare improper this simplifying, yet totally consistent, assumption. To him, it represented an unnecessary, additional hypothesis that violated the purist molecular mentality, which had proven the complete adequacy of the principle of "sensible force at insensible distance" in the analysis of a variety of terrestial phenomena. For Navier to postulate a particular law relating force to displacement was, to Poisson, to adopt the conjectural approach of eighteenth-century philosophers. It was a step backwards.

Navier first derived the equations of motion from the consideration of equilibrium of a single molecule, but then he recast his entire analysis in terms of the variational approach of Lagrange's *Analytique mécanique* in order to obtain an appropriate system of boundary conditions. One deficiency of Poisson's memoir of 1814 lay in his inability to derive a corresponding set of conditions for the plate. He would not resort to Lagrange's method, since it violated, in principle, the purely mechanical molecular approach. So Poisson had ample reason to make objections to Navier's analysis despite its apparent significance. Once again a swift judgment of Navier's endeavors was not forthcoming. Indeed, over a full year passed without any official word from the Academy concerning his two memoirs. Once again Navier remained active.

On the eighteenth of March, 1822, he was invited to read a third memoir, "Sur les lois du mouvement des fluides en ayant égard a l'ad-hésion des molecules." Poisson and Fourier were again chosen to evaluate this new treatise in which Navier employed the molecular model to derive a system of differential equations describing the flow of fluids.[7] Here again was an important paper in theoretical mechanics. Here again Poisson would have seen Navier as a competitor.

But more than Poisson's attitude, it was Cauchy's behavior that finally moved Navier to complain of the Academy's procrastination. The minutes of the Academy of Sciences record that on the thirteenth of September, 1822, Cauchy described his theoretical investigations on the behavior of elastic and fluid media, the same phenomena that Navier had treated.[8] It appeared that Cauchy had taken improper advantage of Navier's work. In a communication to the Academy at its next session, Navier demanded that the long-overdue reviews of his two memoirs on the flexure of elastic plates and on the laws of equilibrium and movement of solid elastic bodies be speedily accomplished.[9] Still, no action was taken.

As unfairly as Navier had been treated, it would be unjust to charge Cauchy with plagiarism. In the first place, though stimulated by Navier's work, Cauchy believed that Navier's approach to the plate flexure problem was lacking. He set himself the task of correcting what appeared to be a conceptual deficiency in the way Navier had treated the internal forces generated within a plate due to its bending and stretching. Navier had assumed that these forces act perpendicular to an imaginary plane cutting across the thickness of the plate. Cauchy thought this inappropriate and unnecessarily restrictive.

In the second place, Cauchy approached the behavior of an elastic solid in a distinctly different way. Whereas Navier's analysis was based on the molecular conceptual scheme promoted by Poisson, Cauchy treated the elastic medium as a continuum, as an infinitely divisible medium capable of supporting internal forces on arbitrarily oriented planes of arbitrarily small area. From his deliberations emerged an analysis "Récherches sur l'équilibre et le mouvement intérieur des corps solides ou fluides, élastiques ou non élastiques." After his September reading, an abstract of this work was sent on to the *Bulletin de la Société Philomatique* and published in its January edition of 1823. In his opening paragraph Cauchy acknowledged only Navier's plate memoir, although the object of his efforts had more to do with Navier's second memoir:

These investigations have been undertaken, occasioned by a memoir published by M. Navier the fourteenth of August, 1820. In order to establish the equation of equilibrium of an elastic plate, the author considered two kinds of forces produced, one due to the extension and contraction, the other kind due to the flexure of this same plate. Furthermore, he assumed in his calculations that both kinds were perpendicular to the lines or to the faces against which they act. It appears to me that these two types of forces could be reduced to one type alone, which uniformly ought to be called tension or pressure, and which has the same nature as hydrostatic pressure exerted by a fluid at rest against the surface of a solid body. Only this new pressure does not always remain perpendicular to the faces that support it, nor [does it remain] the same in all directions at a given point. In developing this idea, I arrive at the following conclusions: . . .[10]

The following conclusions included: a definition of stress, a derivation of the equations of equilibrium in terms of stress, a definition of strain, a derivation of the equations relating strain to the displacement of each point of an elastic body, and a postulated system of stress-strain relationships.[11]

Cauchy had thus formulated, with one swift stroke, most of the essential ingredients of today's theory of elasticity. His work might be used in place of the first three or four chapters of any text on this subject.

Navier's reaction to Cauchy's work appeared in the March issue of the same journal. Navier stressed that the major importance of his first memoir lay not in the rederivation of an accepted plate equation, but rather in the practical solutions obtained from this equation. He then went on to claim priority with respect to the object of his second memoir, namely, a general theory of elasticity.

The memoir of M. Navier of 14 August, 1820, has never been 'published', only some lithograph copies were distributed to several persons. This memoir was referred for examination by the Academy to a commission of which Cauchy is the *rapporteur* . . .

The demonstration of the differential equation of the elastic surface constitutes only the least part of the work contained in the memoir of 14 August, 1820, and the author attaches no importance to it. The special object of this work is the investigation of the flexural behavior of a plate loaded with weights, an investigation based on the integration of this equation, which has been known for a long time.

Besides this memoir of which we speak, its author presented to the Academy of Sciences on the fourteenth of May, 1821, another "Mémoire sur les lois de l'équilibre et du mouvement des corps solides élastiques." His intent was to find the general conditions, expressed in terms of differential equations, upon which the displacement of the points of solid elastic bodies depend. The memoir contains the indefinite equations that apply to the interior points of the solid, as well as the determinate equations that apply to points at the surface.

The aim of the work that M. Cauchy has publically announced, appearing to have the greatest analogy with that of the memoir we are speaking of, renders it important to recall and state the date of this memoir.

Subsequently, Navier did have extracts of both memoirs published in the *Bulletin Société Philomatique*.[13] Finally, although no formal report on these treatises was ever made, his work led to his election, in January, 1824, to the Mechanics Section of the Academy of Sciences.[14]

The treatment of Navier within the Academy bears some similarity to that accorded Sophie Germain: they both had reason to feel that their work was used improperly by others. She had been brought in contact with Navier's work by Fourier, who, a day or so after Navier had read his paper on the flexure of plates to the Academy, sent his own lithograph copy to Sophie Germain.[15]

It seems likely that Navier and Sophie Germain met, though there is no correspondence that proves personal acquaintance. That Navier had some kindly feeling toward her is evidenced in the opening paragraph of the published abstract of his 1820 memoir. It reads:

The interesting experiments of M. Chladni on the vibrations of plates have stimulated the application of the calculus to the laws of movement what are manifest in these experiments: this was the subject of a prize proposed by the first class of the Institute and won by Mademoiselle Germain. The research that was awarded the prize was founded on an ingenious hypothesis, namely, that flexure gave birth, at each point of an elastic plate, to a force proportional to the sum of the inverse values of the two radii of principle curvature. Mademoiselle Germain gave the differential equation of equilibrium and movement of an elastic plane and some integrals of these equations, analogous to those that Euler has given for the elastic lamina.[16]

Comparing this abstract to the introduction of his 1820 lithographed version of his analysis of plates, Navier here attributed a greater significance to Sophie Germain's contribution to plate theory. Now her work was "founded on an ingenious hypothesis"; in the lithograph edition of 1820 only her comparison of particular solutions to the equation Lagrange had derived with Chladni's experimental results had been cited. Here, too, all reference to Lagrange has been dropped. Navier's increased esteem for Sophie Germain arose, most likely, from recognizing in her a fellow-sufferer at the hands of certain members of the Academy. He embraced her as an equal not in mathematics certainly, but in mistreatment.

Yet, this distinction, however obvious to us, could not be obvious to Sophie Germain. She felt herself in the middle of activity in elasticity, working, as with Navier, outside the Academy. She had been wronged by Poisson and he by Poisson and Cauchy.

What sense she might have made of the work going on, however, is unclear. Not only was the work of Navier and Cauchy shifting the interest of mathematicians from the behavior of the elastic plate to that of an arbitrarily shaped elastic solid but, in addition, alternate conceptual schemes for analyzing this behavior were advanced that clearly made Sophie Germain's "ingenious hypotheses" unequal to the task. On the one hand, Cauchy had introduced the concepts of stress and of strain within the context of matter viewed as a continuum – undoubtedly, a perspective foreign to Sophie Germain's experience – while Navier had adopted the molecular approach of her rival Poisson.

She must have felt a bit bewildered, but she had in no way been deterred from continuing her work and her efforts to stay abreast of new developments. The next memoir she wrote attempted further generalization of her research to deal with plates of varying thicknesses, possibly as a way of bridging the gap between the plate and the arbitrarily-shaped solid.

This memoir seems to have been written largely for the purpose of proving herself correct and in the midst of productive work. Its effect was quite the opposite. Although she did address a discrepancy displayed in her analysis, that of Navier, and that of Poisson (each deduced a different expression for the way in which the thickness of the plate appeared in the plate equation), Sophie Germain's argument in support of her own position was clearly inadequate. Furthermore, her analysis of plates of non-uniform thickness is slight and unsuccessful.[17]

Sophie Germain submitted her manuscript to the Academy for approval in 1824. Fourier, acting in his capacity as Perpetual Secretary, sent Sophie Germain a short note informing her that her memoir had been presented at the Academy's meeting of Monday , March 8, and that Laplace, Prony, and Poisson had been chosen to review and report on her work. A favorable assessment would result in a recommendation that her work be published.[18] Fourier, acting as friend, also wrote her the following informal letter.

Mademoiselle,   Friday Morning (12 March 1824)
    I terribly regret not having been able to respond as promptly as I would have wished on the subject of the mathematical memoir you sent me. I have faithfully performed the errand you gave me when you sent it to me. M. Cuvier was charged last Monday with the reading of this correspondence. I asked him to present your memoir and I indicated the purpose of it to him. After the reading, Messieurs Laplace, Prony, and Poisson were named commissioners. I will insist as much as it might be necessary that they make the report you desire. If M. Poisson has designs of showing some opposition to the result of your research, he will not be able to keep from ceding to the authority of experience, whom no one knows better than you to consult. As far as I could determine from your discussion, it seemed to me that you lay completely bare the insufficiency of the theoretical hypothesis from which he desired to deduce the equation of the fourth order, which you have found. I would not be able to contribute to the examination of the memoir without putting aside the immediate affairs I find myself occupied with. All the persons present at the meeting listened to the announcement of your memoir with the greatest interest. The difficulty of the subject, the renown of the authors who have considered it, and your name could not fail to excite attention. Several of us are discussing it at the Academy and at the home of Laplace. I thank you, Mademoiselle, for the new indications of interest you gave me, in occupying yourself with my health and my work. Public discourse is an annoying business, and those persons whose support I esteem the most are those whom I fear most highly as an audience.
    I would have preferred to tell you personally of the presentation of your memoir, and I will take advantage of another occasion to speak to you about it. I am presently detained by much less agreeable affairs.
    Accept, Mademoiselle, with the offer of my thanks, that of my respects.
                                                                      J. Fourier
P.S. The minutes of the meeting that I have drawn up mention the reading of your memoir. You have probably not received the [formal] letter in which I inform you of the names of the commissioners, because it is not customary to make them known until the minutes have been read and adopted.[19]

Clearly, Sophie Germain was afraid of invidious treatment from Poisson and had appealed to Fourier for a fair hearing. He would not tell her –

evidently no one could tell her – that her detailed analysis simply did not constitute first-rate work. She was left without any critique. In fact, no report was ever made on her accomplishment or lack of it. Poisson read her work, probably judged it unworthy of any further serious attention, and passed it along to Prony. Evidence indicates that her manuscript went no further; it remained in Prony's possession, or that of his estate, until July, 1880.[20]

In 1826 she fashioned another memoir and sent it on to the Academy. Essentially, it consisted of an attempt to present a more lucid version of her analysis by introducing several simplifying assumptions.[21] This time, however, she published her memoir first, without submitting it for approval, and then sent it to the Academy. Under these circumstances there was no requirement for review, though Cauchy agreed to make a verbal report. Arrival of her memoir was acknowledged at the Academy's session of 17 July, 1826. Again, Fourier officially expressed the Academy's appreciation, informing her that Cauchy had been designated to review her work.[22]

The day after her memoir was received, Sophie Germain wrote Cauchy the following letter:

Monsieur,   18 July [1826]
   I see with pleasure that my new remarks have been submitted to your judgment. Any other would appear less sure to me.
   I have followed your judgment, by publishing. I would be happy if you would take the trouble of communicating your observations to me. There are in this small work three things that seem to me to be of some importance. First, resolution of this question: from what basis does one derive knowledge of the conditions governing this particular case of the motion of solid elastic bodies? One sees, without it being necessary for me to say it, that the numerous experiments of M. Savart do not bear on this question. They could not be explained without the aid of a more extended theory, such as that which you are already working on. Indeed, if in these experiments one could separate the different layers, which one can conceive the thickness [of a plate] to be composed of, each one would present a particular shape. The thickness would also vary as a result of movement. I am sure of this, and besides, M. Savart has observed it. I have begged M. Ampère to ask you where I could find what you have published on the general case of motion of elastic bodies. He did not do it, and I have been able to have only a cursory reading, insufficient for my instruction. As well as I can recall, from what you have taken the trouble to explain to me yourself, this motion must be considered as being composed of and produced by forces that act in all possible directions; the motion of a surface would present the particular case where the resultant of the forces acting on each molecule would be perpendicular to the different tangential planes. I would be delighted if you were to take up this area of research, and my poor work would take on a real importance in my eyes, if it were to direct your attention there.
   A second consideration on which I would like your opinion is that of mean curvatures. I

have already spoken of this in the first memoir I published. What happens with respect to curvature is what one sees happening in many other phenomena. I wish to say that the actual state of points, equidistant from the positions exhibiting extreme behavior, will give the mean state of the system. It is thus that the temperature of points equidistant of those of maximum and minimum temperature is equal to the mean temperature of the section. In this case it requires an equal quantity of heat to change a given mean temperature. With respect to elastic surfaces, it requires equal forces to change a given mean curvature, so that the curvature of the sphere is always comparable to that of a surface of arbitrary shape. It appeared to me that this remark might be of some use, and it helped me to give a sufficiently simple form to the general equation of surfaces. Finally, this same equation seems incontestable to me, and I would very much like to know what you think of it.

Perhaps you will find it an abuse of your good nature that I add the annoyance of this commentary to that of reading my small memoir. I would have no other excuse but the importance I place on your judgment. I beg you, Monsieur, to accept the assurance of this at the same time as that of my respect.[23]

This letter is revealing in a number of ways. It seems likely that Cauchy had encouraged her to publish in order to relieve the Academy of the embarrassment of having to deal with her memoir. They could not approve work that was so inadequate and trivial, but at the same time they could not treat her as a full professional colleague, as they would any man, by simply rejecting the work. But how was she to know this? Once again, respect for her sex led to such disrespect for her person that she was allowed to appear ridiculous.

She was, furthermore, working under impossible conditions. The letter itself indicates this. Cauchy had given her an oral explanation of a complicated matter, probably in passing at some social gathering, since they were not really friends and there is no indication that he made a real effort to explain his work to her. Much more likely she had him trapped behind a tea table; he was forced to talk a bit of mathematics in order to extricate himself. That she was unacquainted with the essence of his work is clear from her arguing that the "resultant force" acts perpendicularly to the tangential planes of a surface. It sounds absurdly arrogant for her to suggest that he take up this area of research, but she had no way of knowing that. He had published only that one extract of his work in 1823, an extract that she had seen; but, as she indicates in the letter, she was unable to get copies of his full memoirs in order to study them properly. His students at the École des Ponts et Chaussées could learn of his work, but Sophie Germain could not.[24]

It was impossible for one outside the scientific community to take a central part in developments within that community. While Sophie Germain had been working on the competition there was no conflict: she worked alone and nothing was happening elsewhere. As is often the case, the first work done is done alone and on a fairly simple level. This then creates interest among a group of people who, sharing and exchanging ideas and techniques, produce more sophisticated work. The elasticity problem had now moved into a community to which she did not and could not belong, even though she did not know it, and even though that community would not or could not really tell her so. This difficulty existed quite independently of her mathematical inadequacies, though these inadequacies came from the same source, namely, her being a woman and thus excluded from formal education. That she had the native ability and intuition is illustrated by the remarkable amount that she taught herself and by the flashes of insight that appear in her work. Yet, even if she had become superbly proficient in variational techniques and thoroughly at home with physical principles, there is no likelihood that this would have qualified her for entrance into the scientific community. Lack of access to the sessions of the Academy, the inaccessibility of memoirs, and lack of a place in the professional community would still have been determinant. When elasticity became a topic of interest among the professionals, a woman had no place in their midst.

Cauchy acknowledged the receipt of her memoir and letter with the following:

Mademoiselle,   Sceaux-Ponthievre, 23 July, 1826
    I have received the letter that you have done me the honor of writing, along with the memoir that accompanied it. I thank you for having sent me a copy of that work, which I will read with all the care that the importance of the subject and the merit of the author deserve.
    Accept, I beg, the respect with which I am, Mademoiselle, your humble and very obedient servant.
                                                            A. L. Cauchy[25]

But no report on her work was ever made.

The name of Poisson was curiously absent from new work in elasticity after 1814. He could easily ignore Sophie Germain's efforts as the fanciful expressions of an incompetent, but Navier's adoption and adaptation of the molecular mentality in his clear and economical analysis of the behavior of an elastic solid must have caused him some distress. Not until 14 April, 1828, did the Academy hear his response. Perhaps a memoir on the subject in the molecular mode by Cauchy, presented to the Academy in October, 1827, stimulated Poisson to complete his work, or perhaps it was the publication, in full, in 1827 of Navier's short, definitive piece from

1821.[26] Poisson finally produced a massive memoir meant to encompass all previously treated elastic phenomena and some, as yet, unexplained problems, all within the embrace of a true molecular vision. One clear blemish in Poisson's work, appearing in a summary published in the *Annales de Chimie* in midyear, 1828, was his failure to cite Navier's prior accomplishment. His historical introduction reads as if no one other than Poisson had accomplished anything significant in elasticity theory during the first three decades of the nineteenth century.[27]

Navier could scarcely fail to reply. In a note published in the next volume of the same journal, he took issue with the accuracy of Poisson's history:

When M. Poisson read to the Academy this article, in which the author mentions the principal investigations of ancient geometers who addressed the subject that he treats, I was unable to avoid recalling what I myself had accomplished, which was contained in a memoir presented in 1821, published as an extract in the *Bulletin de la SociétéPhilomantique* in 1822, and printed in its entirety in 1827, in volume 7 of the *Mémoires de l'Académie*. M. Poisson responded that the citations relating to my work were recorded in the body of his memoir . . . but I would have been more pleased with the benevolence or justice of the author if, not treating the living less well than the dead (since the immense superiority of the one is no reason to forget the other), he had made this citation in the preamble or extract of his work. In fact, this preamble is known to the Academy, is published in scientific journals, and read by all those who have a taste for the mathematical and physical sciences. The memoir, however, since it contains a sophisticated analysis, reaches only a small number of scientists. Thus, the citation that the author has relegated to the memoir remains almost unknown.[29]

Navier's claim to priority stimulated a reply from Poisson. What had been the subject of behind-the-scenes discussion and intellectual struggle burst out into the pages of the *Annales de chimie* in a polemic ranging over a spectrum of ticklish issues and not just confined to priority in publication.[29]

Navier cited Poisson's deficient analysis of the behavior of the elastic plate, also mentioning Sophie Germain's "ingenious hypothesis." Poisson immediately exposed the confusion inherent in her use of the term "elastic moment" and pointed to an apparent error in Navier's plate boundary conditions. Navier responded that there was no error.[30]

Poisson spoke as though Lagrange had made no substantial contribution to plate theory. Navier described how Lagrange had used Sophie Germain's hypothesis to derive the plate equation. As evidence he made public Lagrange's note to the commission charged with evaluating her first entry. How did this note come into Navier's hands? As a member of the Academy he would have had access to it, but possibly it had been in Sophie Germain's possession and she gave it to Navier.[31]

Much of Poisson's critique of Navier's work concerns the way in which Navier made use of the molecular model. It is a pedantic critique, of questionable importance, motivated by Poisson's sense of authority in molecular matters. All in all, Poisson comes off the worse in this exchange, which signaled the end of the molecular mentality as the only accepted mode for the investigation of elastic phenomena.

Sophie Germain felt compelled to join the fray beyond simply supplying evidence to Navier. Indeed, her response to Poisson's memoir preceded Navier's in the pages of the *Annales de chimie* of 1828.[32] In this "Examen des principes qui peuvent conduire à la connaissance des lois de l'équilibre et du mouvement des solides élastiques," she first described her own endeavors, and recalled that, in 1821, she had criticized the insufficiency of the molecular force hypothesis. Here she claimed priority in this business of exposing the limitations of the molecular mentality. Yet while Navier and Poisson dealt with the details of scientific substance in their exchange, Sophie Germain remained aloof and removed from technical details in her essay. She wrote in a more general philosophical mode, which was free of the glaring insufficiencies of her analytical experiences. In the process she fashioned some insightful remarks on the nature of scientific inquiry:

> Let me be permitted to recall that the object of mathematics is not to investigate the causes that one can assign to natural phenomena. This science would lose both its character and credit if, renouncing the support of general, well-confirmed facts, it sought within the realm of nebulous conjecture, a realm which has always been a fertile source of error, for ways of satisfying the thirst for explanation. With regard to the forces of elasticity, the general, special, and characteristic fact is the tendency of bodies endowed with such forces to reestablish themselves in the shape that an external cause made them lose. This tendency requires that all the molecules of an elastic body tend also to take up again the position that they occupied prior to the action of an external cause that had displaced them.
>
> Such is the fact, the only incontestable fact of elasticity; and if, in order to gain an idea of how this fact makes itself known, one desires to push further, one should fear the introduction into the problem of considerations that are either useless or even entirely foreign to the problem.[33]

Sophie Germain's admonition was directed at those who talked of molecules and sensible forces at insensible distances, those who searched in the domain of nebulous conjecture for causes. She suggested that the weaving together of causal conjectures with mathematics to explain physical phenomena can be deceptive and result only in lowering the status that mathematics rightly deserves. Her brief discourse is reminiscent of the way Laplace once spoke of his predecessors who had postulated various behavioral characteristics and attributes of molecules in search of scientific knowledge.

There is an appealing quality to her remarks. They stand as a refreshingly blunt expression of positivistic methodology. At the same time, her essay reveals the difficulty of applying philosophical principle to scientific practice. The one "fact" basic to Sophie Germain's endeavors provides a valid *definition* of what we mean by an "elastic" body but that is not enough: the development of a *theory* of elasticity requires one to "push further;" her "fact" lacks the richness of a more nebulous conjecture, which is essential to theory building and detailed analysis. Sophie Germain's critique nevertheless possessed merit: the molecular model opened the way to intricate mathematical activity and placed elastic behavior within the context of a broad spectrum of terrestrial phenomena; but it also, as she warned, leads to deception and error. The molecular model, as employed by Navier, by Poisson, and occasionally by Cauchy, is now discredited as a basis for the analysis of the elastic behavior of solids. Essential results derived from this conceptual scheme proved to be wrong.

The efforts of Sophie Germain, of Navier, and of Poisson, however, cannot be discounted in the emergence of a plate theory and a more general theory of elasticity. They were part of a historical process that involved rigorous analysis, competent experiment, ingenious hypotheses, fertile concepts, and the extension of mathematical prowess; a process, also, that involved intellectual prejudice, personal antipathy, political maneuvering, and the use of position for the benefit of friends. Out of all this a theory of elasticity emerged.

CHAPTER TEN

FINAL YEARS

In her last years, Sophie Germain became ill with cancer. No longer able to engage in mathematical studies with her previous energy and persistence, she turned to more general cultural questions.

A most ambitious essay, "Considérations sur l'état des sciences et lettres," was left in an unfinished state.[1] There is no way of knowing what her final plans for it might have been. A short first chapter speculates on the relationship between scientific and artistic intellectual endeavors and a long second chapter traces the history of human intellectual development, the nature of society, drifting off at the end into a survey of the state of literature and, finally, of music.

To some extent this essay was the logical outgrowth of her early habit of writing down *pensées,* abstract thoughts about the nature of things stimulated by her daily life and work.[2] Most were scientific in origin. When she came to write up her thoughts about science and culture in a fully developed and orderly manner, it was natural for her to use her own experience in mathematics as the basis of her ideas.

At the same time, however personal her essay is, it is also rooted in the culture of the time and was published because of the currency of many of the ideas with which it was concerned. The social circles in which Sophie Germain travelled included Auguste Comte. His early work, which provided the foundation for his philosophy, appeared early in the twenties. Hegel visited Paris and was welcomed by the same circle. Refuting Kant was a current philosophical task that Sophie Germain also undertook in the body of her essay. Indeed, it was largely on account of this essay, fitting as it did so nicely into the growth of Positivism, that Sophie Germain's memory was preserved at all. Her nephew published it two years after her death, and forty-five years later H. Stupuy edited a whole book on Sophie Germain, including some of her correspondence as well as the essay, evaluating the latter within the context of Comte's work.[3]

From a twentieth-century perspective, her argument for the similarity of all intellectual endeavor is the most interesting and provocative point in her writing. All such work, she said, from mathematics to the arts, requires both imagination and reason. One seems more apparent than the other in different work, but in all things, reason, which is what we would tend to think of as scientific method, or at least disciplined thought, is necessary for working through and confirming what must be grasped first of all by imagination. Here again, she was speaking from her own experience, her own work in mathematics, where she had known how little value methodological prescriptions were without initial insight. She was also speaking of an experience that is shared by scientists and artists, a fact that is rarely acknowledged.

Sophie Germain's "Considérations sur l'état des sciences et lettres" has a ring of summing-up. In that respect it fits with two memoirs of 1830, one on the curvature of surfaces and another on number theory, both summarizing work she had done much earlier.[4] All these writings reveal a desire to record all the things she had left to say, to complete her work before she died.

The memoir on curvature has a history behind it that once again includes Gauss. In her memoirs of 1821 and 1826 Sophie Germain had introduced the notion of a referent sphere as a measure of the curvature of a surface. (The radius of the *sphere* equals the mean curvature at a point of the surface in the same way that, at each point of a plane curve, one can identify a referent *circle* whose radius defines its curvature.)[5] Just as the elastic force for the beam is measured by the radius of a referent circle, so, too, the elastic force of the surface is measured by the radius of the referent sphere. The 1831 paper on the curvature of surfaces, her final elaboration on this concept, may have been provoked by the appearance of a memoir on the same subject by Gauss.

In 1827, Gauss published his *Disquisitiones generales circa superficies curvas.*[6] In this work he defined a measure of the curvature at any point of a surface, as Sophie Germain had. He, too, used an equivalent sphere as a device to assist in the analysis, however, his definition of curvature differed from hers. To her, curvature was measured by the *sum* of the principal curvatures, while he defined it as the *product* of these two quantities. This was not a matter of arbitrary choice; both had their own reasons for their definitions. For Sophie Germain, analogy to the curvature of a plane curve, and more importantly, Lagrange's equation determined her position. For Gauss, the more geometric consequences of his definition were determinant.

Sophie Germain did not learn of Gauss's work until early in 1829, and then only by chance, in the course of a conversation with Monsieur Bader, a student of Gauss's, who visited Paris that year. In the following letter, Sophie Germain revealed to Gauss her thoughts about the curvature of surfaces.

Monsieur,

I take advantage of the return of your learned disciple Monsieur Bader to your service by thanking you for your kind remembrance and also by sending you some new copies of my memoirs.

I have read, with great pleasure, your memoir on biquadratic residues, which this young scientist has given me on your behalf. It suffices to sustain my appetite for arithmetic research, reminding me, Monsieur, that this has, at other times, procured the honor of receiving several letters from you. Do believe that I heartily regret being deprived for such a long time of those learned communications to which I have never ceased to attach the greatest value.

In chatting with Monsieur Bader about the current subject of my study, I provided him with the occasion to speak to me, and subsequently, to show me, the learned memoir in which you compare the curvature of surfaces to that of the sphere. (I would very much have liked to have been able to keep this memoir. I have given it [to him] to take back but very reluctantly because I do not know where to find it again.)

I cannot tell you, Monsieur, how astonished, and at the same time, how satisfied I was in learning that a renowned mathematician, almost simultaneously, had the idea of an analogy that seems to me so rational that I neither understood how no one had thought of it sooner, nor how no one has wished to give any attention to date to what I have already published in this regard.

Apart from the superiority that characterizes all that flows from your scholarly pen, there exists an essential difference between your research and mine. Yours, Monsieur, is entirely geometric; mine, on the contrary, is in a way mechanical and only involves geometry to the extent necessary to establish (in the case where there is a connection) the identity of the forces whose expression has been the almost exclusive object of my research for such a long time.

I am going to attempt, Monsieur, to give you a view of the way in which I have been led to compare the curvature of a surface to that of a certain sphere. You know, Monsieur, that in naming $r \: r'$ as the two radii of principal curvature of an originally flat elastic surface, the shape of which is changed by the effect of an external cause, I have affirmed that the force with which the surface tends to return to in natural shape is proportional to $1/r + 1/r'$. I learned (by the way, I only obtained knowledge of Poisson's objections to my theory in a circuitous and more or less uncertain manner) that one of the commissioners, relying on Euler's discussion of the infinite number of different curves obtained from the intersections of different planes passing through a given point of the surface, had thought that I had not sufficiently established the choice of principal curvatures and that I was not justified in saying that they give the measure of the curvature of this surface at the chosen point. In searching to rebut this objection, I remarked that in taking the sum of the inverses of the radii of curvature of two of the curves of normal intersection, it would suffice to choose those that are contained in two mutually perpendicular planes in order to obtain a constant quantity equal to the sum of the inverses of the two radii of principal curvature. Further-more, I remarked that if the two planes of intersection chosen make an angle of 45 degrees with those that contain the lines of principal curvature, the radii of curvature of the curves contained in the chosen planes are equal to each other and to the radius of a sphere that cuts the surface in such a way that two of the quadrants will be below and the two others above the surface. This sphere is what I have named *the sphere of mean curvature*.

I am in the process of proving, in a superior way relative to what I have published previously in this regard, that whatever be the shape of the element of the surface, that is to say, whatever be the manner in which the curvature of the element is distributed about the point of tangency, the force that would be employed to destroy the curvature of this element remains constant . . .

I regret being deprived of the advantage that I would enjoy, as will Bader, from your wise consideration – what he takes for granted does not astonish me, but it is for me an object of envy. Independent of what I could learn from you, I regret further not being able to submit to your judgment a multitude of ideas that I have not published and that would take too long to write out.

Wishing at least, Monsieur, that you remember me and agree to the assurance of my profound respect.

Your servant,

Paris 28 March, 1829   Sophie German[7]

Evidently Gauss had suggested to Bader that he visit Sophie Germain and deliver to her a copy of his latest paper on number theory when he went to Paris. It seems likely that Gauss also knew of her work in elasticity, though if he did he was not sufficiently impressed by it to mention it to Bader: it was only in the course of her conversation with Bader that Sophie Germain learned that Gauss had himself written a memoir on the curvature of surfaces. Yet Gauss had been favorably impressed by her ability in number theory, and he had mentioned her work occasionally to others before. She remained someone to whom he could recommend a travelling student.

This letter from Sophie Germain to Gauss is, in some respects, the most overtly complaining letter in all her correspondence, revealing a rather clear perception of her position in the scientific community as well as a twinge of despair that that position amply justified. On the matter of memoirs, she had not even known that Gauss had written on curvature; and she still did not have a copy of the memoir. The one that Bader had shown her she returned with no assurance that she could get another. Although she had sent Gauss copies of her own memoirs, with a clear plea for a copy of his in return, he evidently did not respond (evident only because we would expect a letter of thanks from her).

Beyond the matter of published work that, as it happened, was far from publicly available, we find Sophie Germain acknowledging and lamenting how isolated she is from other people. She stated her envy of

Bader in returning home to Gauss. Was she possibly asking for an invitation? Would it have been possible for her to have made such a visit? We don't know. We only know that she made no such trip. Whatever she may or may not have been hinting at, she did not openly beg for a letter, a personal communication from Gauss. Nor did he send one.

The substance of her letter, discussing her approach to curvature in relation to his, suggests that she spent rather little time with his memoir before returning it. She saw in his memoir what was most similar to her own work, namely, his use of a referent sphere, and seemed quite unaware of his definition of Gaussean curvature, though she did have a sense of some important difference between their work. She made the same kind of mistake that she had made in her letter to Cauchy in 1821 after she had just had a brief look at his memoir ("not sufficient for my instruction"): she saw what appeared similar to her own work and missed what he was really doing. It is possible, of course, to say that she simply couldn't understand the complexity and sophistication of what was being done. To some degree this is true, since there were concepts that she clearly had difficulty with, e.g., the Lagrangian definition of "elastic moment." Yet it belittles her intellect unfairly to blame all her difficulty on that. In fact, whenever one reads something new, one notices first what is already familiar. It is only through rereading and careful study that one sees where an author diverges from the preconceptions of the reader. Sophie Germain was not afforded an opportunity for extended study, hence her response to Gauss's memoir on curvature was impressionistic and showed a lack of understanding of his work; instead she used her impression of what he was doing to stimulate further thought of her own.

Despite her remark to Gauss that she was thinking of these things, Sophie Germain was no longer working towards publication and an active role in the development of elasticity theory. It was political circumstance that led to her final memoir on the elastic behavior of surfaces.

Libri wrote in his obituary of her:

When the revolution of July [1830] broke out, she took refuge in her study as she had during that of '89; it was during the week of fighting that, taking up and developing further some old ideas, she wrote her "Mémoire sur la Courbure des Surfaces," which appeared in the *Annales* of M. Crelle of Berlin.[8]

There is something touching as well as amusing in the image of Sophie Germain, fifty-four years old, already sick with cancer and within a year of her death, responding to the world around her as she had at the age of thirteen. It is likely that the analogy was one that she had drawn herself, rather than an invention of her biographer: Libri was an elegant and aristocratic Italian whose writing is full of graceful and rhetorical flourishes, most of which fit neatly into a rather mannered prose style. Such an observation as this, however, seems more likely the product of a conversation with a real person than a literary invention.

Libri knew Sophie Germain on an easy social level during the last years of her life. He was twenty-seven years her junior, and had met her in 1824, when he came to Paris from the University of Pisa, where he had been named professor of mathematics the preceding year. Their friendship arose from a mutual interest in number theory: Libri, when still in his teens and hearing of the recent prize for work on Fermat's Last Theorem, had studied the work of Legendre and Gauss. In 1820, he published a memoir entitled "Memoria supra la teoria dei numeri," a copy of which he sent to the French Academy, which duly reported on it.[9] Thus when he came to Paris he was welcomed by the scientific community as a young and promising mathematician and there he met Sophie Germain.

Their friendship began with mathematics, but quickly extended beyond those bounds. Sophie Germain, at the age of forty-eight, established however uncomfortably in the older circle of Parisian scientific society, was in a position to play patroness to an aspiring young Italian. Libri was in a position to enjoy and benefit from this relationship. Beyond practicalities, however, they seemed to have enjoyed each other's company.[10]

The closing paragraphs of Libri's brief biographical sketch gives us a contemporary picture of Sophie Germain during the latter part of her life:

Mademoiselle Germain was not devoted only to mathematics; she, in addition, mastered different areas of knowledge, any one of which would have established the reputation of a woman. She was very competent in the natural sciences. Then too, she had learned Latin on her own, not for its own sake, since in her view languages were only instruments of study, but in order to enable her to understand diverse works, notably those of Newton and Euler. In addition, among her papers have been found some very subtle philosophical reflections, for she was actively occupied with metaphysics, which she claimed was the source of the true philosophical spirit. She thought very little of diverse philosophical systems, which she viewed as the literary productions of superior intellects.

Her conversation was of a completely unique nature. Its striking characteristics were an unfailing ability for recognizing a basic idea in an instant and arriving at its final consequences, jumping over all intermediaries; humor, whose gracious and light form belied always an exact and profound thought; an ability, which derived from the variety of her studies, to reconcile similarities between the physical order and the moral order, which she regarded as subject to the same laws. If one adds to this an unfailing benevolence, which caused her always to think of others before herself one will sense what must have been her charm.

This forgetting of self she displayed in all her activities. It was evident in science, which she cultivated with a complete denial of self, never thinking of the advantages that success procures. She rejoiced even when she saw her ideas made fruitful on occasion by other persons who adopted them. She repeatedly stated that it matters little who first arrives at an idea; rather what is significant is how far that idea can go. She said, happily, that her ideas had produced their fruit for science while not yielding anything for her reputation – which she scorned and amusingly called the glory of the bourgeois, the small place we occupy in the mind of others.

This noble character she also displayed in her actions, actions always marked by the stamp of virtue, which she said she cherished as a mathematical truth. Since she could not conceive that one could love the ideas of one kind of order without loving those of another, the ideas of justice or virtue were, following her thinking, ideas of order that the mind ought to adopt, even when the heart did not cherish them.

Such was this superior woman, who of all who pursued mathematical studies the farthest, the only one, to our knowledge, who has made real progress. The theory of sound and indeterminate analysis will keep her name alive for a long time.

This side of Sophie Germain – the accomplished woman at ease in social situations, with her own unique personality – is one we have seen little of. Through her work she has appeared primarily in a position requiring struggle and reaction.

Such difficulties began when she was a child, seeking an education that she couldn't quite get. First there were the arguments with her family about her private studies, as she tried to learn the basics of mathematics from books alone. Then there were her efforts to benefit from schools that she could not think to attend in person. We know very little about these attempts, about what she tried and where, but certainly there were plans and schemes leading up to her emergence as "Leblanc" in a notebook submitted to Lagrange.

Failure to conceal the results of her studies led to the one period in her life when the tension between her desires and the expectations of people around her relaxed somewhat. She was young, female and talented in mathematics by *men's* standards. For these reasons she attracted the attention and curiosity of some of the people she most wanted to know. Even at this time, however, there was friction between her interests and intentions on the one hand and those of her admirers on the other. On the most obvious level it became clear in the case of Lalande's unacceptable gallantry. More seriously, it was inherent in the way her acquaintanceships developed, e.g., the letter among her papers replying to questions about paradoxes. This shows the willingness of her correspondent to answer her questions, but that is all. What Sophie Germain so clearly needed was an orderly education both in basic mathematics and in judging the nature of mathematical questions; she needed criticism as well as encouragement, discipline as well as indulgence. She appeared to her new friends as a prodigy to be marveled at, rather than as an adept student to be taught.

The most helpful relationship she had during the early part of her career was surely that with Gauss, a relationship that blossomed largely by chance. Gauss had few with whom he could share number theory at the time, and was motivated for purposes related to his career, to pursue problems of lesser interest to himself. He was pleased, therefore, with an interesting correspondent whom he had every reason, at the beginning, to assume was a man. By the time he learned her sex it made no difference, since they never met and their communication was already established.

Sophie Germain's mature life was filled with struggle. This occurred almost by chance, certainly not through her intention. The issues involved in this struggle are much clearer now than they were at the time. Some of them arose simply from her sex: when she was young, mathematicians could not see that she required training more than admiration. When she was older, they could not see that she required extended discourse, and the opportunity to participate alongside them in their work more than formal recognition. Sophie Germain was most likely quite aware of what she lacked, but was not in a position to frame her case effectively, or argue it cogently: self-awareness was just beginning to be phrased in economic terms; it would be a long time before it could be put in sexual terms.

Yet in one respect Sophie Germain could and did express quite clearly what the basic technical issue of her struggle within the professional community was about, namely whether her hypothesis or her mathematical technique were at fault in her analysis of the behavior of the elastic plate. It was convenient, to be sure, for her to overlook the mathematical objections, but she also knew that she was speaking to a real issue, namely, the sanctity of the molecular mentality espoused by Laplace, Biot, and Poisson.

Surely Poisson, beholden to a molecular vision as a way of explaining all terrestrial phenomena, would not have accepted her basic hypothesis as anything more than fortunate conjecture. Indeed, in his 1814 memoir, he showed how an alternative hypothesis – that the elastic forces be assumed proportional to the *difference* of the principal curvatures – yielded the same equation for the behavior of the interior points of the plate.[12] If Sophie Germain had been able to do the mathematics that Lagrange never revealed but which Poisson certainly had worked through, the issue would have been as she had interpreted it, i.e., a matter of justification of her hypothesis. As it was, her awkward handling of variational technique allowed Poisson to ignore her and to dismiss her efforts as eager, but incompetent.

In number theory she was on firmer ground. Both Gauss and Legendre recognized her talent in this domain. She had flirted with a host of enticing questions before she settled on Fermat's last theorem for her most concentrated and productive work. But in number theory, as in applied mathematics, her work reveals competence, not the energy of potential genius. Genius displays an ability to recognize the form of a whole where others see only disconnected questions, a talent for choosing and defining probems that are somehow intuited, and only in retrospect clearly seen, as crucial ingredients of a general theory. Competence remains content to play with odds and ends. Fermat's Theorem was ignored by Gauss; it did not warrant his effort. Legendre, and Sophie Germain, on the contrary, were attracted to this quite special problem. To Gauss we attribute the production of seminal papers and the setting of a foundation for theory. To Sophie Germain we attribute but one significant theorem. The rest of her work appears as the meandering motions of a light craft in uncharted waters. Gauss appears to be in control of his destiny; she somewhat adrift.

Their work on the curvature of surfaces serves as another example. Sophie Germain was led to her definition of mean curvature and its properties as a result of a struggle with a particular physical problem. Gauss, too, was no doubt motivated by practical concerns, for he was involved in a geodesic survey at the time. Yet Gauss stepped back from the practical dimension and found in his activity a basis for abstraction and generalization, and a fertile source for the generation of related mathematical questions. We grant to Gauss not only the discovery of the measure of curvature bearing his name, but also significant accomplishments in setting the stage for conformal mapping, and non-Euclidian geometry. In contrast, Sophie Germain described an analogous physical phenomenon, that of steady state temperature distribution in a solid body, and suggested that it was subsumed by her notion of mean curvature.

Sophie Germain would have been the first to acknowledge that she lacked true genius. Yet competence, too, has its use and rights, and her goal in the world of mathematics was to be able to work among her peers, and to justly admire her superiors. Despite all the skirmishing, she seems never to have taken any pleasure in the struggle itself. In a revealing letter to Gauss in 1809 she wrote:

I do not pretend to fathom the profundity of your research. I sense that my intellect is far removed from yours, although our tasks are similar, since I, as you, have a great predilection for arithmetic problems. I find this part of science susceptible to a particular kind of elegance, which is not attained in the mathematical-physical sciences. It appears that in everything the interest of ideas is in inverse proportion to the usefulness they have in practice. This is not surprising when we consider that the human intellect, when working for its own satisfaction, should encounter the greatest intellectual beauties rather than when guided by an external motive. . . .[13]

This letter was written just before she found herself lured into studying the practical plate problem, with all the accompanying appeal of a prize and public acknowledgment. Her first love, however, was pure abstraction, pure intellection unmixed with worldly interests. This predilection was supported by her financial independence and her professional obscurity. It lent her character an honest detachment from practical considerations. This in turn provided a context for Libri's characterization of her social presence as intelligent, quick witted, gentle, and indulgent.

The last words we have from her own pen were written to Libri two months before her death:

18 April, 1831

Monsieur, I am more afflicted than astonished by what you tell me concerning your present situation. I see you disquieted, the disposition of your spirit far removed from the exclusive love of the sciences, which would have made you happy. You say that you have written me twice since your departure. I have not received one letter except for that written at your departure from Mr. Maurice's. I have delivered to Madame Cauchy the letter you gave me. . . .

My health is in a frightful state. A prompt death would be a relief to me, because I suffer from unbelievable pain, which leaves me not a moment's rest. I wanted to read at least the third volume of De LaCroix, but I cannot. I remain shut up. I see neither M. LeGendre nor my other friends, except for St. Amant [Lherbette], who is always concerned about you, and my sister. I am told that my condition is not desperate, but I am warned of long suffering. I have received from M. Crelle the issue that contains one of your memoirs and mine. He asks for news of you. I told him of your departure from Geneva. As for me, I told him that it would be absolutely impossible for me to profit from his good will, at least not for a long time. There is definitely a fate hanging over all mathematicians – your unhappy preoccupation, that of Cauchy, the death of M. Fourier. Finally, that student Galois, who in spite of his impertinence, displays a good disposition, has managed to be expelled from the Ecole Normale. He is without fortune and his mother has very little. He continues the injurious behavior of which he gave you a sample after your best lecture at the Academy. The poor woman has left her house, leaving him enough to live in a mediocre way, and has been forced to place herself as a *dame de compagnie*. One hears that he is becoming totally insane and I believe it.

I address this letter to Marseille, as you informed Madame Renaud. If you can give me news of yourself, do so immediately, I pray you. I desire at the same time some news of your mother. Although I do not have the honor of knowing her, it is impossible for me not to be concerned with the worries that she must feel.

Accept the assurance of the constant interest that your talents have inspired in me.

Sophie Germain[14]

This harried and unhappy letter may well reflect the tone of the last period of Sophie Germain's life. Thinking about herself, she could only see illness, loneliness, and incapacity. Thinking about others gave her little relief: somehow she had managed to make friends with difficulty and misfortune. Except for the tenacity with which she had pursued her interests, Sophie Germain had never stepped beyond the bounds of decorum, but in her later life she seems to have been drawn to those who did. The student Galois was destined to die in a duel a few years hence, leaving to posterity an epochal piece of work on the theory of equations hastily written up the night before he died. Libri himself had been stripped of his Italian citizenship and sent into exile. Thus as she lay on her death bed, she worried about her friends, and wished she had the strength to study.

The death certificate for Sophie Germain records her merely as a *rentier*, i.e., a person of private means.

NOTES

CHAPTER ONE: INTRODUCTION

[1] At the turn of the century, Laplace (1749–1827) was probably the most influential person in matters political as well as scientific within the First Class of the Institute. Lagrange (1736–1813) could claim as much respect for his accomplishments in mathematics but was less involved in non-scientific affairs. Legendre (1752–1833), a third mathematician, less brilliant than the first two but also a member of the First Class, had on occasion been snubbed by Laplace. A good, brief summary of Laplace's work is found in Andoyer, H.: 1922, *L'Oeuvre scientifique de Laplace*, Payot, Paris. Some of the details of a priority conflict between Laplace and Legendre are also given there. S. D. Poisson (1781–1840), Joseph Fourier (1768–1830), A. L. Cauchy (1789–1857) and C. L. M. H. Navier (1785–1836), all mathematicians of a younger generation, were, in that first decade of the nineteenth century, striving for recognition. See Gillespie, C. C.; (ed.): 1972, *Dictionary of Scientific Biography*, Charles Scribner and Sons, New York. For Poisson, see also Arago, D. F. J.: 1854–62, 'Notices biographiques,' *Oeuvres*, 17 vols., Gide et J. Baudry, Paris, 2, pp. 593–671.

[2] Todhunter, J., and Pearson, K.: 1886, *A History of the Theory of Elasticity and of the Strength of Materials*, At the University Press, Cambridge.

[3] Navier, C. L. M. H.: 1827, 'Mémoire sur les lois de l'équilibre et du mouvement des corps solides élastiques', *Mem. Acad. Sci.* **7**, pp. 375–394. The importance of this memoir, read to the Academy of Science on the 14th May, 1821, has been argued in Todhunter, J., and Pearson, K., *op. cit.*, p. 138 and in Timoshenko, S. P.: 1953, *History of the Strength of Materials*, McGraw-Hill, New York.

[4] Stress is a measure of internal force, like pressure, and is measured in units of force per unit of area. Strain is a measure of deformation, e.g., extensional strain may be taken as the ratio of change in length of a line element (due to deformation) to the original length of this line element.

[5] It is curious, however, that Navier's one-constant theory and contemporary continuum theory differ only slightly in their predictions of the elastic behavior of isotropic solids. If one of the two constants in modern theory is set at 1/4, one obtains Navier's system of equations. For most structural materials, that constant is quite close to 1/4.

[6] Volume 37 of the *Annales de Chimie* (1828) contains Navier's initial remarks on Poisson's work. Their polemic is described in Chapter 9 below. (See footnotes 28 and 29 of that chapter.)

[7] The sincerity of Napoleon's patronage of science is called into question by L. Peace Williams in Williams, L. P.: 1956, 'Science, Education and Napoleon I', *ISIS* **47**, pp. 369–382. A more positive evaluation of the Emperor's influence is presented by M. Crosland in Chapter 1 of Crosland, M.: 1976, *The Society of Arcueil*, Harvard University Press, Cambridge, Massachussetts. See also Lacour-Gayel, G.: 1911, *Bonaparte, membre de l'Institut*, Gauthier-Villars, Paris, and Barral, C.: 1889, *Histoire des sciences sous Napoléon Bonaparte*, A. Savine, Paris.

[8] Chapter 3 of Crosland, *op. cit.*, contains a concise description of the structure and functions of the First Class of the Institute. See also Maindron, E.: 1888, *L'Académie des*

sciences, F. Alcan, Paris, Gauja, P.: 1934, *L'Académie des sciences de l'Institut de France,* Gauthier-Villars, Paris, and Hahn, R.: 1971, *The Anatomy of a Scientific Institution: The Paris Academy of Sciences, 1666–1808,* University of California Press, Berkeley and Los Angeles.

[9] Lagrange's work on the vibration of strings dates to 1762. See Lagrange, J. L., 'Recherches sur la nature et la propagation du son', in *Oeuvres,* 14 vols., 1867–1892, Gauthier-Villars, Paris, 1, pp. 39–148. For a review of eighteenth-century developments in the mechanics of structures see Truesdell, C., 'The Rational Mechanics of Flexible or Elastic Bodies, 1638–1788', in *Leonhardi Euleri Opera Omnia,* Ser. 2, 30 vols., 1912–1964, 11 (pt. 2).

[10] Chladni, E. F. F.: 1809, *Traité d'Acoustique,* Chez Coucier, Paris. This review was included as an appendix in Chladni's book. The 'Rapport adopté par la Classe des Sciences Mathématiques et Physiques et par celle des Beaux-Arts, dans les séances du 13 février et de mars 1809, sur l'ouvrage de M. Chladni relatif à la théorie du son', occupies pages 362–375.

[11] *Ibid.*

[12] In 1808, the interest these patterns held was largely centered on the search for mathematical analysis – a problem in applied mathematics, since it involved physical phenomena, rather than purely abstract reasoning. The fascination of the problem as it appeared then is quite different from that motivating studies of vibrating plates today. Now modes of vibration are of interest and importance in engineering problems. Thus, e.g., sand patterns on a resonating triangular plate, simulating a sweptback wing, appeared in the May 1953 *Journal of the Aeronautical Sciences.* Gustafson, Stokey, and Zorowski: 1953, 'An Experimental Study of Natural vibrations of Cantilevered Triangular Plates', *Journ. Aero. Sci.* **20**, p. 331. Further, an article appearing in the *Journal of the AIAA* in February, 1972, displayed modes of a circular plate supported at three points on its perimeter that were obtained from a computer analysis. Chi, C.: 1972, 'Modes of Vibration in a Circular Plate with Three Simple Support Points', *Jour. AIAA* **10**, p. 142. The patterns found in these modern studies are the same as some of those displayed in Chladni's book.

[13] "Rapport adopté . . ." *op. cit.,* p. 357.

[14] Carl Friedrich Gauss (1777–1855), one of Western civilization's grandest mathematicians, lived in Brunswick and later in Göttingen, but never visited Paris. See Dunnington, G. W.: 1955, *Carl Friedrich Gauss, Titan of Science,* Hafner, New York. A more concise treatment of his genius and his achievements is provided by Hall, T.: 1970, *Carl Friedrich Gauss* (trans. A. Froderberg), MIT Press, Cambridge, Mass.

[15] Todhunter and Pearson, *op. cit.,* p. 156.

[16] To Cauchy we owe the introduction of the concept of stress and the framing of elastic theory within a continuum context. See Cauchy, A. L.: 1823, 'Recherches sur l'équilibre et le mouvement intérieur des corps solides, élastiques ou inélastiques', *Soc. Philom. Bull.* (1823), pp. 9–13. This is only a brief review of Cauchy's memoir that was read to the Academy of Sciences in September, 1822. The memoir itself is not readily available; however, in a later treatise published in his *Exercices de mathématiques,* 'Sur les équations qui expriment les conditions d'équilibre ou les lois du mouvement intérieur d'un corps solide, élastique ou non-élastique', Cauchy state that most of what appears in this article was originally reported on in 1822. See Cauchy, A. L.: 1828, *Exercises de mathématiques, Année 1828,* in *Oeuvres complètes,* Ser. 2, 13 vols., Gauthier-Villars, Paris, 8, p. 215.

CHAPTER TWO: SOPHIE GERMAIN

[1] The genealogical facts derive in the main from Stupuy. H., (ed.): 1896, *Sophie Germain, Oeuvres Philosophiques*, (Nouv. Ed.), Paris. (Hereafter cited as Stupuy.) This one volume, containing the editor's study of the significance of Sophie Germain's life and work, a collection of a few dozen letters written by or addressed to Sophie Germain, and a reprinting of her philosophical writings, originally appeared in 1879. With his first edition, H. Stupuy, member of the Municipal Council of Paris and of the Departmental Council for Primary Instruction, triggered a surge of interest in Sophie Germain as woman, mathematician and philosopher. In 1888, the Ecole de la rue de Jouy, a school for young women established in 1882, was renamed in her honor. The school still thrives.

[2] A more thorough and trustworthy study of the Germain family was accomplished by Madame Dufour and appeared as a *Supplément au bulletin de l'Association Amicale des Anciennes Elèves de l'Ecole Municipale Supérieure Sophie Germain*, (1932). A copy of this can be found in the Archives of the Academy of Sciences, Paris.

[3] Jacques-Amant Lherbette was responsible for a first printing of Sophie Germain's essay. Lherbette, J., (ed.): 1833, *Considerations générales sur l'état des sciences et des lettres*, Lachevardière, Paris.

[4] After her mother died in 1823, Sophie Germain moved again, this time to the Left Bank. A commemorative plaque hangs by the entrance of this modest place on the rue de Savoie.

[5] Montucla, M.: 1758, *Histoire des mathématiques*, 2 vol., C. A. Jombert, Paris.

[6] Libri, G.: 1832, 'Notice sur Mlle. Sophie Germain', *Journal des débats*, 18 May. This obituary article, written by a mathematician of some fame, and friend of Sophie Germain, appeared in Lherbette's edition of the *Considérations générales . . ., op. cit.*, and has since served as the major source of information about Sophie Germain.

[7] Libri, *op. cit.*

[8] *Encyclopedia Britanica*, 27th ed. (1959) *s.v.* 'Sophie Germain'.

[9] Libri, *op. cit.*

[10] The few details surviving about LeBlanc are located in *Registre des élèves, Ecole Polytechnique*, vol. 2, held at the Archives of that institution. He was born the 5th of July, 1775, had brown hair, brown eyes, and was 5 ft. 9 in. tall, and resided at 21 rue des Marais, Faubourg St. Germain.

[11] One finds in a folder of assorted administrative records held in the Archives of the Ecoles des Ponts et Chaussées a 'Liste des Elèves de l'Ecole Polytechnique admis a l'Ecole Nationale des Ponts et Chaussées et des supplémentaires' dated 1 Nivoise, an 6. The eighth name on the list of admissions, just above Poinsot, is LeBlanc. A pen stroke runs through his name and the word *mort* appears in explanation.

[12] Libri, *op. cit.*

[13] Bibliothèque Nationale, MS. Fr., 9118, 'Corres. de Mlle Sophie Germain'. All but two items in this collection appear in Stupuy, *op. cit.* By 1797, Sophie Germain no doubt would have read Cousin's book on the calculus. Cousin, J.: 1777, *Leçons sur le calcul différentiel et le calcul intégral*, C. A. Jombert, Paris.

[14] Bibliothèque Nationale, MS. Fr., 9118. Published in Henry, C.: 1879, 'Les manuscrits de Sophie Germain – documents nouveaux', *Revue Phil.* **8**, p. 623.

[15] Joseph-Jerôme LeFrançais de Lalande, 1732–1807, ". . . was extremely well-known during his lifetime partly because of the enormous bulk of his writings and partly because of his love for the limelight." So states T. Harkins in Gillespie, C. C., (ed.): 1972, *Dictionary of*

Scientific Biography, Charles Scribner and Sons, New York, *s.v.* 'Lalande'.

[16] Bibliothèque Nationale, MS. Fr. (N.A.), 4073. Published in Henry, *op. cit.*, p. 635.

[17] The first edition of Lalande's *Astronomie des dames* was published in 1785. It was reprinted in 1795, again in 1806 and in 1820 – the last version accompanying Fontenelle's *Entretiens sur la pluralité des mondes*.

[18] Bibliothèque Nationale, MS. Fr. 9118. Published in Stupuy, *op. cit.*, p. 250.

[19] *Ibid.*

[20] *Ibid.*

[21] *Ibid.*

[22] de Villoison, J.: 1802, 'Vers de Jean-Baptiste-Gaspard d'Ansse de Villoison, membre de l'Institut national de France, pour le jour de la naissance du célèbre astronome Jérôme de Lalande (le 11 juillet)', *Magasin encyclopédique* **1**, pp. 238-240.

[23] Lalande's niece – actually the wife of the grandson of Lalande's uncle – occasionally performed some calculations for her 'uncle.'

[24] Figures II.1 and II.4 are reproduced from Stupuy, *op. cit.* The bust, (Figure II.2) the work of Zacharie Astruc, today stands in the courtyard of l'Ecole Sophie Germain. The sketch (Figure II.4) was done by Mme. Silvain Dufour, the school's first director.

CHAPTER THREE: RESPECTFULLY YOURS, GAUSS

[1] Libri, G.: 1832, 'Notice sur Mlle. Sophie Germain', in Lherbette, (ed.), *Considérations générales sur l'état des sciences et des lettres*, Lachevardière, Paris, (1833).

[2] Gauss, C. F.: 1801, *Disquisitiones Arithmeticae*, G. Fleischer, Leipzig. A French translation was published in 1807; Poullet-DeLisle, (trans.): 1807, *Recherches arithmétiques*, Chez Courcier, Paris. The first English translation appeared in 1966; Clarke, A. A. (trans.): 1966, *Disquisitiones Arithmeticae*, Yale University Press, New Haven, Conn.

[3] Froderberg, A. (trans.): 1970, *Carl Friedrich Gauss _ A Biography by Tord Hall*, MIT Press, Cambridge, Mass. The full significance of Gauss' work was not immediately appreciated. Legendre reported on a single gem contained in this treatise – probably Gauss' proof that whenever $N = 2^{2^n} + 1$ is a prime number, one can construct, by Euclidean methods, a regular polygon of N sides – at a meeting of the First class of the Institute. The minutes of that meeting suggest that Gauss's name had yet to become a household word. "Citizen Legendre reported on a geometric discovery made in Germany by M. Charles Frederic Bruce, (sic) of Brunswick, published in his work entitled, *Disquisitiones Arithmeticae*, Leipzig, 1801." Hendaye (ed.): 1910–22, *Procès-verbaux des séances de l'Académie des Sciences*, 10 vols., Imprimé de l'Observatoire d'Abbadia, 2, p. 457. (Hereafter referred to as *P.V.*) Another bit of evidence suggests that Legendre, and probably Lagrange, found the *Disquisitiones* . . . heavy going. (They would most likely have been the two academicians charged with reviewing Gauss's book.) In Delambre, J.: 1810, *Rapport historique sur les progrès des sciences mathématiques depuis 1789* . . ., de l'Imprimerie impériale, Paris, the perpetual secretary of the class for Mathematical Sciences states "M. Gauss has treated in an entirely new way this entire theory (of numbers), in a uniquely remarkable work, about which it is impossible for us to give any summary, since all of it is new, even its language and its notation."

[4] Facsimile reproduction in Boncompagni, B.: 1880, *Cinq Lettres de Sophie Germain à Charles-Frederic Gauss*, B. Boncompagni, Berlin. Published also in *Archiv der Math.*

Phys., Lit. Bericht, 1880, Leipzig, 259, and in Stupuy, *op. cit.*, while a copy, verified by E. Shering, is held at the Bibliothèque de l'Institut, MS. 2031. The actual letter is held at the Universitätsbibliothek at Göttingen. Sophie Germain's draft is contained in the Bibliothèque Nationale, MS. Fr. 9118.

[5] *Ibid.*

[6] *Ibid.*

[7] *Ibid.*

[8] Gauss's letter, following a six-month interlude, is dated 16 June, 1805, (Bibliothèque Nationale, MS. Fr. 9118). Also published in Stupuy, *op. cit.*, p. 258, but with the date in error (16 June, 1806).

[9] LeBlanc wrote to Gauss on 21 July, 1805. See *Cinq lettres . . .*, *op. cit.*; *Archiv der Math. Phys.*, *op. cit.*; or Bibliothèque de l'Institut, MS. 2031. Gauss responded with a brief note on 20 August 1805. See Bibliothèque Nationale, MS. Fr., 9118 or Stupuy, *op. cit.*, p. 261. The next item in this exchange, from LeBlanc to Gauss, is dated 16 Nov., 1805. See *Cinq lettres . . .*, *op. cit.*; *Archiv der Math. Phys.*, *Lit. Bericht*, 1881, Leipzig, 261, Bibliothèque de l'Institut, MS 2031, or Stupuy, *op. cit.*, p. 263. Sophie Germain's draft is in the Bibliothèque Nationale, MS. Fr. 9118.

[10] LeBlanc to Gauss, 21 July, 1805, (see above).

[11] Some may consider it a minor mystery why Gauss devoted nearly all of his time and energy to astronomy during this first decade of the nineteenth century when he expressed so much interest in, and displayed such genius for, arithmetic research. The most likely explanation, crudely put, is that there was not much of a market for the sublime truths of higher arithmetic. In addition, there was intense interest in the application of mathematics to the prevailing problems of astronomy. Determination of the orbital elements of the four minor planets, Ceres, Pallas, Juno, and Vesta, discovered in the period 1802–1807, particularly attracted the attention of the continental geometers. Sophie Germain was well aware of Gauss's interest in these problems. In his short letter of 20 August, 1805, he informed her that he was occupied ". . . with perfecting several new methods of calculating planetary perturbations. These and the methods I have used to calculate the elliptical elements of the various new planets will probably furnish the material for my next work." It was Gauss's accomplishments and contributions to the resolution of these problems that brought him both economic and professional reward. His election to a position as Correspondent to the Geometry Section of the First Class of the Institute in 1804, his acquisition of a professorship in astronomy at Göttingen in 1807, and his winning of the Lalande medal (of 500 francs in value) in 1809, awarded by the First Class of the Institute for his treatise *Theory of Planets and Means of Determining their Orbits*, were all the result of his work in astronomy and celestial mechanics. Perhaps the myth of free and unfettered research by great minds into areas of intense personal interest may never have contained any more truth than it appears to claim today.

[12] LeBlanc to Gauss, 21 Nov., 1804, (see note 4, above).

[13] LeBlanc to Gauss, 16 Nov., 1805, (see note 9, above).

[14] Gauss to LeBlanc, 16 June, 1805, (see note 8, above).

[15] Joseph-Marie Pernety, general and senator, born in Lyon, 1766, died in Paris, 1856.

[16] This Madame Lalande is the wife of the nephew of the astronomer Lalande whom we encountered in the second chapter.

[17] Bibliothèque Nationale, MS Fr. 9118. Published in Stupuy, *op. cit.*, p. 266.

[18] *Ibid.*, p. 269.

[19] *Cinq lettres . . .*, *op. cit.* Also *Archiv der Math. Phys.*, 1881, *op. cit.* and Stupuy, *op. cit.*, p. 271. A copy, verified by E. Shering, is in the Bibliothèque de l'Institut, MS. 2031.

[20] See Gauss, C. F.: 1863–1929, *Werke*, Teubner, Leipzig, 10, pp. 70–74. Also in Stupuy, *op. cit.*, p. 274.

[21] *Cinq lettres . . .*, *op. cit.* Also in *Archiv der Math. Phys.*, 1881, *op. cit.*

[22] Stupuy, *op. cit.*, p. 283.

[23] On 21 November, 1807, Gauss had arrived in Göttingen to take up his duties at the University.

[24] Quoted in Dunnington, G. W.: 1955, *Carl Friedrich Gauss, Titan of Science*, Hafner Pub. Co., New York, p. 67.

[25] Gauss, *Werke*, *op. cit.* 10, p. 566.

[26] *Ibid.*, p. 74.

[27] Quoted in Dunnington, *op. cit.*, p. 68.

[28] *Ibid.*, p. 69.

[29] Stupuy, *op. cit.*, p. 285.

CHAPTER FOUR: SETTING THE PRIZE

[1] See Crosland, M.: 1967, *The Society of Arcueil*, Harvard University Press, Cambridge, Mass.

[2] Quoted in Maindron, E.: 1881, *Les fondations du prix de l'Académie des Sciences*, Gauthier-Villars, Paris, p. 55.

[3] *Ibid.*, p. 53.

[4] *Ibid.*, p. 54.

[5] *Ibid.*, p. 69. Another annual prize ". . . for the most interesting observation or the most useful memoir contributing to progress in astronomy . . ." was established that same year at the request of Lalande. The value of the award varied from year to year and depended on the interest earned on the 10,000 francs Lalande had placed in trust for this purpose. See, again, Maindron, 1881, p. 66.

[6] Reception of Chladni's book on acoustics was acknowledged at the First Class's meeting of 24 October, 1808. See *P.V. vol. 4, p. 119*. His request for an examination of his clavi-cylinder was noted at the next week's meeting on 31 October.

[7] Burckhardt, an observational astronomer, was elected to the Astronomy Section of the First Class in 1804. That same year the Bibliothèque Germanique was established and Burckhardt became one of its four editors, responsible for reviewing publications in the mathematical sciences. *P.V.* vol. 3, p. 173.

[8] *P.V.* vol. 4, p. 139 (Session of 28 November, 1808).

[9] This commission's report on the clavi-cylinder was read on 19 December, 1808. See *P.V.* vol. 4, p. 147. Their review of Chladni's work on acoustics, "Rapport adopté par la Classse des Sciences mathématiques et physiques et par celle des Beaux-Arts . . .", was reprinted as an appendix to the French translation of Chladni's book. Chladni, E.: 1809, *Traité d'acoustique*, Chez Coucier, Paris, pp. 362–375.

[10] *P.V.* vol. 4, (session of 6 February, 1809).

[11] *P.V.* vol. 4, p. 162. (session of 13 February, 1809).

[12] The announcement of this special contest in mathematics was published as an appendix to Chladni's *Traité d'acoustique*, *op. cit.*, pp. 353–357.

[13] A substantial summary of S. D. Poisson's (1781–1840) life and work, authored by Arago, and read at a public session of the Académie des Sciences, 16 December, 1850, appears in Arago, D. F. J.: 1854–62, 'Notices biographiques', *Oeuvres*, 17 vols., Gide et J. Baudry, Paris, 2, pp. 593–671.

[14] Jean-Baptiste Biot (1774–1862), a graduate of the Ecole Polytechnique in 1797, was appointed assistant astronomer at the Bureau des Longitudes by Laplace in 1806, and elected to the mathematics section of the First Class of the Institute in 1803. Laplace's influence on the nature and direction of Biot's scientific work was considerable. See Gillespie, C. C., (ed.): 1972, *Dictionary of Scientific Biography*, Charles Scribner and Sons, New York, *s.v.* 'Biot'. Dominique François Jean Arago (1786–1853) also attended the Ecole Polytechnique, also held a position at the Bureau des Longitudes – that of secretary – and was elected to the astronomy section of the First Class in 1809. The circumstances of his election, which depended very much on Laplace's interest in his candidacy, are summarized in Crosland, *op. cit.*, in a section entitled 'The Election of Members of the Arcueil Group to the First Class and to the Académie des Sciences' (pp. 161–168).

[15] Poisson's 'Sur la pluralité des integrales dans le calcul des différences' was read before the First Class at its meeting of 8 December, 1800. Lacroix and Legendre, charged with reviewing this work, praised it highly and recommended that it be published in the *Recueil des savants étrangers*. Arago, *op. cit.*, p. 609. It was published in the *Journal de l'Ecole Polytechnique*, vol. 4, (11 cah.), pp. 173–181.

[16] Two memoirs in particular elicited an enthusiastic response from the First Class. In his 'Théorie du son', read to the Institute on 17 August, 1807, Poisson employed the accepted mathematical formulation for the propagation of sound through air and, introducing an ingenious technique – method of images – deduced the way in which sound reflects from a barrier and also the acoustic field produced within an ellipse with a sound source located at one focus. He then went on to consider in a more speculative way a problem that had bothered Laplace and others before him: the unexplained discrepancy between the experimentally observed speed of propagation of a sound wave through air and the speed deduced theoretically by Newton. As Biot relates in reviewing Poisson's effort, Laplace had previously offered "an ingenious explanation of this difficulty by attributing the acceleration of sound to changes in temperature experienced by the particles of air as they condense and dilate." Poisson was not able to express Laplace's qualitative explanation analytically by giving it mathematical form but he did manage to show that other factors and conditions would not provide an explanation. His efforts were crowned by the approval of the Class. *P.V.*, vol. 3, p. 594. With the reading of his next memoir, in June 1808, Poisson achieved full stature as a mathematician. Again Biot, along with Laplace and Lagrange, reviewed his presentation. In this treatise, "Sur les inégalités séculaires des moyens mouvements des planètes," Poisson addressed a problem that had also been considered by both Laplace and Lagrange, namely the stability of planetary orbits about the sun. Poisson's analysis extended and completed their solutions of this problem and thereby insured the lifetime of a stable sun-planet system effectively for eternity. Biot's opening remarks in his review of this memoir suggested that something of major significance had been accomplished, for he began by extolling the glory of mathematics applied to astronomy: "It has given society a calendar to regulate work and pleasure; it has dissipated the foolish terrors with which celestial phenomena have previously struck men in times of ignorance. It has given the navigator sure and facile procedures for determining his position *en route* on the high seas; it has given nations means of precisely fixing the events of their history; and in

place of preserving epochs by vague souvenirs or perishable monuments, it has enabled them, so to speak, to trace them in the sky . . ." And, in closing, Biot praised Poisson's significance: ". . . one sees that the memoire of M. Poisson includes several results of prime importance for astronomy. It required great wisdom, a combination of most delicate analysis and profound knowledge of the complete mathematical theory of celestial motion. The Class notes with interest the constant progress of the talent of this young geometer. It views with pleasure his attention to the theory of the system of the world and it continues to encourage him to devote himself to this kind of analysis, which is the light of astronomy and its primary instrument." *P.V.* vol. 3, p. 89. (Session of 16 August, 1808).

[17] Poisson's talents, as evidenced in his work, were analytical, rather than speculative, nor did they extend to experimental familiarity with phenomena, as Biot or Arago could claim. That he was unsuited for experimental work was evident early in his career when he was apprenticed to an uncle at Fontainebleau to learn the surgeon's trade. To introduce him to the practice of bleeding, his uncle gave him a lancet with which to prick the nerves of a cabbage leaf. The task proved too much for him. Later Poisson recalled, "My hand was so unstable that I never managed to touch those accursed nerves, all enmeshed as they were when I took aim at them." Later, upon his entrance into the Ecole Polytechnique in 1798, Poisson "handled the drawing pen with such awkwardness that they excused him from all graphics work, assuming that he would not enter public service and that his true career would be in the sciences." Arago, *op. cit.*, pp. 595–599.

[18] There is some slight suggestion from the election of Arago to the Astronomy section in 1809 that Poisson's friends were already considering the possibility of his gaining a seat in another section. Arago's election was virtually uncontested: he received 47 votes, Nouet, also on the list of nominees, received 1 vote. But Poisson, who was not on the list of nominees, received 4 votes (Laplace, Berthollet, Biot, Gay-Lussac?). Possibly these ballots were cast for him in the hope of setting a precedent for future nominations to the Astronomy section although until that time it had been defined as a section of observational astronomy. See *P.V.* vol. 4, p. 253.

[19] Etienne Louis Malus (1775–1812) spent a good bit of his life in active service with the army, achieving the rank of major. A graduate of the Ecole Polytechnique, his major scientific achievements lay in the field of optics. His 1808 discovery of the polarization of reflected light was in large part responsible for his election to the First Class in 1810. See *Dictionary of Scientific Biography*, *op. cit.*, *s.v.*, 'Malus'.

[20] Arago, *op. cit.*, p. 603.

[21] This 'note' appears as an appendix to Laplace, P. S.: 1809, 'Mémoire sur le mouvement de la lumière dans les milieux diaphanes', *Mém. de l'Inst.*, *Académie des Sciences* 10, pp. 326–342.

[22] The commission responsible for this decision included once again Laplace, along with Delambre, Burckhardt, Lagrange and Arago.

CHAPTER FIVE: THE ONE ENTRY

[1] Libri, G.: 1832, 'Notice sur Mlle. Sophie Germain', in Lherbette, (ed.), *Considérations générales sur l'état des sciences et des lettres*, Lachevardière, Paris, 1833, p. 13.

[2] Lagrange, J. L.: 1788, *Mécanique analytique*, Chez la Veuve Desaint, Paris.

[3] Euler, L.: 1779, '*Investigatio Motuum quibus laminae et virgae elasticae contremiscunt*',

Acta Acad. Sci., *Petrop.* 1, pp. 103–161. Leonhard Euler (1707–1783), a Swiss mathematician, is recognized, like Gauss, as one of the finest minds to have worked in both pure and applied mathematics. Euler wrote most of his memoirs in Latin – a language in which Sophie Germain had no formal training.

[4] Lagrange, J. L.: 1788, *op. cit.*, 'Avertissement de la première édition.'

[5] Germain, S.: 1821, *Recherches sur la théorie des surfaces élastiques*, Mme. Ve. Courcier, Paris, 1821, (Avertissement). It is unlikely that Sophie Germain witnessed Chladni's experiments in the company of Laplace and Napoleon. More probably, Chladni performed before small, informal gatherings of members of the First Class. Lagrange could have invited Sophie Germain to one of these.

[6] Euler, L.: 1779, *op. cit.*

[7] *Ibid.*, p. 264.

[8] *Ibid.*, p. 265.

[9] A manuscript containing Sophie Germain's analysis of Euler's investigation of the vibration of a beam supported at an interior point is found in the Bibliothèque Nationale, MS. Fr. 9115 f. 155–194. It carries the title 'Remarques sur le mémoire d'Euler: Investigatio motuum quibus laminae et virgae elasticae constremiscunt', in *Acta Acad. Petrop Ann.* (1779) p. 103 *et seq.*'

[10] Legendre is referring to the pagination of Euler, L.: 1779, *op. cit.*

[11] Bibliothèque Nationale, MS. Fr. 9118. Also in Stupuy, *op. cit.*, p. 287. Stupuy consistently erred in the designation of the trigonometric function cotangent. He wrote 'cos' where Legendre wrote 'cot'.

[12] In her 'Remarques sur le mémoire d'Euler . . .', Sophie Germain deduced that in order to obtain solutions which may be interpreted in terms of Euler's Case V, the two equations

$$\text{tang}(1 - \lambda)\omega = \frac{e^{(1-\lambda)\omega} - e^{-(1-\lambda)\omega}}{e^{(1-\lambda)\omega} - e^{-(1-\lambda)\omega}} \text{ and } \text{tang}\lambda\omega = \frac{e^{\lambda\omega} - e^{-\lambda\omega}}{e^{\lambda\omega} + e^{-\lambda\omega}}$$

must be simultaneously satisfied by ω. For most values of λ, She showed that this is impossible.

[13] Bibliothèque Nationale, MS. Fr. 9118. Also in Stupuy, *op. cit.*, p. 291. Legendre missed the point that Sophie Germain had made with respect to solutions of the second kind. She is correct, Legendre in error.

[14] *Ibid.*, p. 295.

[15] Sophie Germain's 1811 entry is held at the Archives de l'Académie des Sciences. A draft is contained in MS. 2381, Bibliothèque de l'Institut.

[16] Bibliothèque Nationale, MS. Fr. 9118. Also in Stupuy, *op. cit.*, p. 298.

[17] *Ibid.*, p. 299.

[18] Bibliothèque Nationale, MS. Fr. 9114. Also in Stupuy, *op. cit.*

[19] Bibliothèque Nationale, MS. Fr. 9118. Also in Stupuy, *op. cit.*, p. 300. Legendre urged Sophie Germain to consult Lagrange's analysis of the membrane. He was, in effect, telling her how Lagrange had obtained the differential equation for the plate starting from her hypothesis (that the elastic force is proportional to $(1/r)+(1/r')$), then applying the techniques of the variational calculus. Lagrange's adoption of Sophie Germain's expression for the elastic force and derivation, of what has since become a familiar equation governing the bending of plates, borders on serendipity. (See note 25, ch. 6).

[20] Sophie Germain, 1811 entry, *op. cit.*

[21] One sees that I take $V((1/r) + (1/r'))$ for the moment of this force. It would take too long

to discuss here those considerations that led me to choose this function of osculatory radii. But it is easy to see that even if the expression for elastic moment ought to contain other functions of these radii than this one I have adopted (such as products or powers of these same quantities), the results, when applied to this problem, would not change since (as is clear in what follows), $(1/r) \cdot (1/r')$, for example, may always be neglected with respect to $((1/r) + (1/r'))$. *Ibid.*

[22] As we shall see, in Chapter 7 below, this note was made public in the course of a polemic involving Poisson and Navier. See Navier, 1828, 'Remarques sur l'article de M. Poisson', *Annal. de chimie* 39, p. 149. The original is held at the Bibliothèque Nationale, MS. Fr. 9118. How it ended up among Sophie Germain's papers is unclear.

[23] The actual letter sent to Legendre is missing, perhaps destroyed. Sophie Germain's draft is found in the Bibliothèque Nationale, MS. Fr. (Nouv. Acq.) 5166.

[24] *Institut de France. Ire Classe. Travaux divers*, vol. 2, no. 24, (1811–1816), (Public Session of 6 January, 1812).

[25] An interesting aspect of the variational approach to the plate-bending problem is that if one assumes, as Sophie Germain did, that the elastic force is proportional to $V((1/r) + (1/r'))$ one obtains, as Langrange did, the correct equation governing the motion of the interior points, but some of the equations governing the behavior at the plate's edges, obtained in this way, are in error. (See note 25, Chapter 6). To Kirchoff (1827–1888) we owe the resolution of these difficulties. Kirchoff, G. R.: 1850, 'Ueber das Gleichgewicht und die Bewegung einer elastischen Scheibe', *Crelles Journal* 40.

[26] Bibliothèque Nationale, MS. Fr. (Nouv. Acq.) 5166.

[27] . . . "sum of the moments of elastic forces that act throughout the plate." Sophie Germain, 1813 Entry, Archives de l'Acad. des Sciences, p. 2.

[28] Bibliothèque Nationale, MS. Fr. 9118. Also in Stupuy, *op. cit.*, p. 303.

[29] *Institut de France. Ire Classe. Travaux divers, op. cit.* (Public Session of 3 January 1814).

CHAPTER SIX: THE MOLECULAR MENTALITY

[1] Poisson, S. D.: 1814, 'Mémoire sur les surfaces élastiques', *Mém. de l'Inst. 1812*, (pt. 2) pp. 167–226. This volume was not published until 1816.

[2] *P.V.*, vol. 5, p. 385 (Session of 1 Aug., 1814); p. 386 (Session of 8 Aug., 1814).

[3] Laplace, P. S.: 1796, *Exposition du système du monde*, Impr. du Cercle-Social, Paris, 2. p. 188.

[4] In his *Opticks*, in a last 'query', Newton had suggested that the refraction of light, capillarity, and chemical combinations could be explained on the basis of forces acting amongst corpuscles. In the *Principia*, Newton analyzed the way in which the pressures and volume of a gas are related proceeding from the assumption that the force acting between any two gas corpuscles is inversely proportional to the distance separating them.

[5] The word 'molecule' as used by Laplace should not be construed in modern terms; its meaning can only be understood within the context of his work. Early in his career, Laplace writes 'molecule' where we would write 'differential element,' e.g. ". . . the molecule dM is equal to a rectangular parallelepiped whose dimensions are dr, $r(dp)$, and $r(\sin p)dq$. . . ." Laplace, P. S.: 1785, 'Théorie des attractions des spheroïdes et de la figure des planetes', *Histoire et mémoires de l'Académie royale des Sciences de Paris*, p. 117. Later, 'molecule' came to mean an isolated particle or corpuscle having a physical identity of its own, e.g.,

". . . I assume that the molecules of gas are at such distances apart that their mutual attraction is insensible. . . ." Laplace, P. S.: 1822, 'Considérations sur l'attraction des corps sphériques et sur la répulsion des fluides élastiques', *Journ. de Phys.* 95, p. 85. Certainly when he conjectures that the shape of a molecule may modify its attraction for, and repulsion of, other molecules, we are required to interpret 'molecule' as corpuscle.

[6] Laplace, P. S.: 1796, *op. cit.*, p. 196.

[7] *Ibid.*, p. 197. At another time, Laplace claimed ". . . we see that the effect of the shape of the molecule must decrease much faster than (gravitational) attraction does as the distances between molecules increase. This applies to astronomical phenomena that depend on the shape of a planet, e.g., the phenomenon of precession of the equinoxes; for this effect decreases with the cube of distance, while the attraction decreases only in inverse proportion to the square of distance." Laplace, *Traité de mécanique céleste*, Book 10, 2nd suppl. in *Oeuvres complètes*, 14 vols., Gauthier-Villars, Paris. 1878–1912, 4, p. 488. The appeal of a force mechanism dependent on the inverse-square law of gravitational attraction but modified by the shape of the molecule was strong. It would enable the reduction of all terrestrial phenomena to the same basic, general law that governed all celestial phenomena. For an example of Laplace's analysis of the resultant force field in the vicinity of a non-spherical body see Laplace, P. S., 'Du developpement en série des attractions des sphéroïdes quelconques", *Traité de mécanique céleste*, Book 3, in *Oeuvres complètes, op. cit.*, 2, p. 24.

[8] Laplace's attitude towards those who would introduce *ad hoc* hypotheses in their quests for explanation is revealed at a later date in the following terms: ". . . I venture to believe that this (molecular hypothesis) will throw great light on the theory of (chemical) affinities for what I advance is founded on mathematical reasoning and not on vague and precarious considerations that ought to be firmly banished from natural philosophy; unless, imitating Newton in his *Opticks*, we offer them merely as conjecture designed to guide us to further research." Boskovitch's elaboration of an inter-molecular force scheme in his *Theory of Natural Philosophy* would have qualified as conjecture worthy of expulsion from the realm of natural philosophy in Laplace's view. Boskovitch, R. J.: 1758, *Philosophiae naturalis Theoria reducta ad unicam legem virum in natura existentium*, Vienna. Yet it has been fashionable to locate the roots of the Laplacian molecular mentality in his work. See Timoshenko, S. P., 1953, *History of Strength of Materials*, McGraw-Hill, New York, p. 104, and Thomson, W., and Tait, P. G.: 1962, *Principles of Mechanics and Dynamics*, pt. 2, Dover, New York, p. 214. Merz's evaluation of the impact of Boskovitch's work is probably the most accurate. "Whilst this treatise represents in general a view largely held by continental philosophers of nature, it does not contain any new mathematical methods such as the 'Principia' contained before and Laplace's 'Mécanique céleste' later. . . . In fact, it is more a metaphysical than an exact treatise, and as such has exerted no lasting beneficial influence on the progress of Science. . . . In France the book seems to have been little appreciated . . .". Merz, J. J.: 1907, *History of European Thought in the Nineteenth Century*, 3rd edn., Blackwood and Sons, London, 1, p. 358. That a French translation of the *Philosophiae naturalis* . . . was initiated in 1779 but never completed tends to support Merz's thesis.

[9] Laplace, P. S., 1796, *op. cit.*, p. 198.

[10] Meetings of this elite group were held at the homes of Laplace and Berthollet, neighbors at Arcueil, a suburb of Paris. An account of the activities of this society is found in Crosland,

M.: 1967, *The Society of Arcueil*, Harvard University Press, Cambridge, Mass. Berthollet's concern was chemical phenomena. Laplace viewed these too as susceptible to molecular analysis; ". . . this law of nature [of sensible forces at insensible distances] is the source of chemical affinities; like gravity, it is not arrested at the surfaces of bodies, but penetrates them, acting beyond the point of contact, but at imperceptible distances. Upon this depends the influence of mass in chemical.phenomena, or the capacity for saturation, whose effects have been so beautifully developed by M. Berthollet." Laplace continued with a discussion of how the shape of molecules, heat and other factors might effect diverse chemical phenomena. He concluded this discussion as follows: ". . . All these phenomena depend upon the shape of the elementary molecules, upon the laws of their attractive force, and perhaps upon other forces yet unknown. Our ignorance of these facts and their extreme complexity do not permit us to reduce the results to mathematical analysis." Chemistry was to remain Berthollet's forte. Laplace remained aloof from these complexities.

[11] Laplace, P. S.: 1822, *op. cit.* 'Chaleur' in Laplace's view behaves as if it too were constituted of molecules; in this case of mutually repelling molecules. In the first edition of his *Exposition du système du monde* (vol. 1, p. 163), Laplace described the phase change of water in the following way: ". . . whatever its nature, heat dilates bodies; it transforms them from solids to fluids and from fluids to vapors. . . Water produced in the melting of ice and vapors by the boiling of water absorbs, at its moment of formation, a great quantity of heat which reappears in the return of vapors to the liquid state . . . If there were no atmospheric pressure, water would change to a vapor; but this pressure balances the repulsive force communicated to fluid molecules by heat and keeps the water in the form of a liquid until the amount of heat becomes so great that its repulsive force overcomes the pressure of the atmosphere. At this instant, water begins to boil and is transformed into vapor."

[12] Discussion of Laplace's use of this molecular conception of heat is found in Finn, B. S.: 1964, 'Laplace and the speed of sound', *ISIS* 55, p. 7, and Kuhn, T. S., 1958, 'The caloric theory of adiabatic compression', *ISIS* 49, p. 132.

[13] Laplace, P. S.: 1822, *op. cit.*, p. 90.

[14] Poisson, S. D., 1829, 'Mémoire sur l'equilibre et le mouvement des corps élastiques', *Mém. Acad. Sci.* 8, pp. 357–570, 623–627.

[15] 'Note' appended to: Laplace, P. S.: 1809, 'Mémoire sur le mouvement de la lumiere dans les milieux diaphanes', *Mém. de l'Inst., Académie des Sciences* 10, pp. 300–325. This 'note', occupying pages 326–342 of this same volume, contains an interesting elaboration of the potential usefulness of the molecular conceptual scheme in analyzing the conduction of heat in solids as well as the elastic behavior of beams. Stimulated by the then-recent innovative work of Fourier on the first of these phenomena and by the dramatic experiments of Chladni on the vibration of plates, Laplace here was staking out a claim for the primacy of the molecular mentality as a basis for the analysis of all physical phenomena. He even showed how the principle of virtual velocities, the basis of Lagrange's *Mécanique analytique*, may be deduced from the interaction of molecules.'

[16] Poisson, S. D.: 1814, *op. cit.*, p. 192.

[17] Laplace's complete analysis of the refraction of light is found in Laplace, P. S.: 'Des refractions astronomiques', *Traité de mécanique céleste*, Book 10, in *Oeuvres complètes, op. cit.*, 4, pp. 233–273.

[18] *Ibid.*, p. 238.

[19] A. C. Clairaut, a mid-eighteenth century savant, had attempted an analysis of capillary

phenomena using the concept of attractive forces acting between the molecules of a capillary tube and molecules of a fluid. Clairaut's approach clearly provided a basis for Laplace's own thinking and Laplace acknowledged the contributions of his predecessor and liaison to Newton in his own work. Laplace was probably responsible for the appearance in 1808 of a second edition of Clairaut's *Théorie de la figure de la terre*. This work, originally published in 1743, contained Clairaut's investigation of the elevation of fluids in capillary tubes. Here Clairaut assumed that "corpuscular attraction is only sensible at a very small distance" but, in Laplace's view, not small enough. For Clairaut had thought that the action of the walls of the capillary tube acted sensibly upon a thin column of fluid imagined lying along the axis of the tube. One consequence of this assumption was that the resultant force of attraction between the tube and the fluid depended on the diameter of the tube. This, in turn, meant that one needed to know how this attraction varied with distance in order to complete the analysis. By negating this assumption, Laplace was able to press forward to obtain the law of ascension of fluids. His work on capillarity appeared in several publications over the period 1804–1809, in 1812 and again in 1819. See Laplace, *Traité de mécanique céleste*. 'Supplément au Livre X' in his *Oeuvres, op. cit.* Poisson's theory of capillarity appeared in 1831. *Poisson, S. D.: 1831, 'Nouvelle théorie de l'action capillaire', Annal. de chimie* 46, pp. 16–70.

[20] Laplace points out that ". . . the integration of this second order partial differential equation resists all known methods", Laplace, P. S.: 1808, *Système du monde*, (3rd ed.), Courcier, Paris, p. 220.

[21] *Ibid.*, p. 235.

[22] This third edition of the *Systéme du monde* contains a convenient summary of Laplace's accomplishments in analyzing capillary phenomena. It is interesting too, that, whereas his discussion of the possibilitiers for molecular analysis in the first edition of this work occupies only the last few pages of a chapter entitled 'Réflexions sur la loi de la pesanteur universelle', by 1808 the proven potential of this conceptual scheme in dealing with refraction and capillarity merited more substantial discussion. An entire section of the third edition is devoted to 'De l'attraction moléculaire'.

[23] This equation was published in 1814 in the form of an abstract which Sophie Germain no doubt read that year. Poisson, S. D.: 1814, 'Extrait de mon memoire sur les surfaces élastiques', *Bull. des Sci. Soc. Philom.*, pp. 47–52.

[24] Application of Poisson's method to the analysis of a beam reveals the inadequacies of his analysis of the elastic behavior of a plate.

[25] Poisson, S. D.: 1814, 'Extrait . . .', *op. cit.* In this summary of his analysis, Poisson described how, since the setting of the prize five years before, only one memoir had been received that was worthy of the attention of the First Class. He stated ". . . at the beginning of this memoir, the unknown author posed, without sufficient proof or, as a matter of fact, without any demonstration, an equation which is precisely our equation . . ." He then related how the author's comparison of particular integrals of this equation with the experimental results of Chladni merited an honourable mention. Poisson viewed the anonymous author's generation of the plate equation as little more than the result of fortuitous conjecture. He had reasons for this opinion that were independent of the inadequacy of Sophie Germain's demonstration; in the same article, he claimed that an alternative hypothesis – that the elastic force is proportional to the difference of the principal curvatures – yields the same equation. In the full version of his memoir (Poisson, S. D.: 1814, "Mémoire . . ." *op. cit.*), which was not published until 1816, he claimed that

application of Lagrange's variational method to *any* expression of the form

$$((1/r) + (1/r'))^2 \pm C((1/r)\cdot(1/r')).$$

where C is an arbitrary constant, produces the plate equation. In the light of this discovery, Poisson, quite rightly, placed little value on Sophie Germain's special hypothesis, $(C = 0)$.
[26] Poisson, S. D.: 1816, 'Extrait d'une memoire sur les surfaces élastiques', *Corres. l'Ecole Polyt.* 3, p. 154.
[27] For a discussion of the political as well as scientific dimensions of the development and demise of the Laplacian molecular scheme see Fox, R.: 1970, 'The Rise and Fall of Laplacian Physics', *Historical Studies in the Physical Sciences* 4, 89–136. Also Fox, R.: 1970, 'The Laplacian Programme for Physics', *Buletin de la Academica Nacional de Ciencao de la Republica Argentina* 48, 429–437. E. Frankel has authored another article focusing on the adoption of the wave theory of light over the corpuscular system; Frankel, E.: 1976, 'Corpuscular Optics and the Wave Theory of Light: The Science and Politics of a Revolution in Physics', *Social Studies of Sciences* 6, 141–184.

CHAPTER SEVEN: AN AWARD WITH RESERVATIONS

[1] Biot, J. B.: 1817, 'Mémoires de la Classe des Sciences Mathématiques et Physiques de l'Institut de France, année 1814; seconde partie, imprimée en 1816; . . .', *Journal des savants* (March 1817), pp. 143–151. Biot composed this piece after Sophie Germain had been awarded the prize (Jan. 1816). To him she appeared as ". . . the only person of her sex who has profoundly penetrated the field of mathematics – even when compared to Mme. du Châtelet, because in this case there was no one like Clairaut [to tutor her]." Daughter of a baron, the Marquise du Châtelet (1706–1749) had a rich and varied social as well as scientific career. Married, mother of three, Voltaire's mistress, under the tutelage of Clairaut she accomplished the sole French translation of Newton's *Principia*.
[2] "I have truly regretted not knowing of the memoir of M. Poisson. I have wasted, in awaiting its publication, a period of time that has been precious to me. I might even have entirely renounced the research that I have the honor of submitting to the judgment of the Class if I had not learned . . . that the [nonlinear] equation obtained from a hypothesis different from what I have proposed, is also obtained from mine. Indeed, each day I see new reasons to regard my hypothesis as incontestable; and yet the respect for the authority of M. Poisson deprived me of the courage to submit to the calculus a principle that I did not then foresee must yield the equation published by this skilful geometer." Sophie Germain's third entry is held at the Archives, Académie des Sciences, Dossier 'Germain'. A draft of the first section of this paper is located in the Bibliothèque de l'Institut de France, MS. 2381, Dossier V.
[3] Poisson's complete analysis of vibrating surfaces was published in 1816, approximately a year after the deadline for the third contest. See Poisson, S. D.: 1814, 'Mémoire sur les surfaces élastiques', *Mém. de l'Inst.*, 1812 (pt. 2).
[4] "It is with regret that I find myself forced to terminate my research here. I know how much it leaves to be desired. But it could possibly take me several years to assemble, for cylindrical surfaces alone, a mass of data yielding results that would have the degree of certainty that has been obtained for plane surfaces by means of M. Chladni's numerous experiments." Sophie Germain, Third Entry, *op. cit.*

[5] *Institut de France, Ire Classe, Travaux divers*, vol. 1, (1811–1816) (Public Session of 8 January, 1816). Sophie Germain did not see fit to participate in this public event. A newspaper of the time reported that "The Institute's Class of Mathematical and Physical Sciences today held its public session before a very large crowd that doubtlessly had been attracted by a desire to see a new kind of virtuosity, Mlle. Sophie Germain, who was to have been awarded the *lames élastique* prize. Public attention has been fooled: this woman did not show up to receive an award which no one of her sex had, up until then, ever won." *Journal des débats*, 9 January, 1816.

[6] Bibliothèque Nationale, MS. Fr. 9114. Also in Stupuy, *op. cit.*, p. 307. Since Poisson's response to this letter is dated 15 January 1816, and the Class did not hear Poisson's report on the results of the third contest until its meeting of 26 December, 1815, it is clear that Sophie Germain's letter was written and delivered in January 1816.

[7] Bibliothèque Nationale, MS. Fr. 9118. Also in Stupuy, *op. cit.*, p. 309. J. Noël Hallé (1754–1822) was a doctor and possibly friend of the Germain family. Contrary to Sophie Germain's claim that her hypothesis yielded his non-linear equation, Poisson states that this is not the case. He provides no explanation, however.

CHAPTER EIGHT: PUBLICATION

[1] *P.V.* vol. 5, p. 595. (Session of 26 December, 1815). Legendre, Laplace, Poisson, Delambre, and Lacroix comprised the commission which set the subject of this competition.

[2] For a full description of Sophie Germain's contribution see Edwards, H. M.: 1977, *Fermat's Last Theorem*, Springer-Verlag, New York, pp. 61–65.

[3] Legendre, Laplace, and Poisson were members of all the various commissions elected to set this prize for the years 1818 and 1820 and to judge any entries submitted in competition. Legendre, having a special interest in number theory, was no doubt the most forceful advocate for a continuation of the first contest after no worthy memoirs were received. Even in 1820, when it was decided not to extend the competition another two years, a new prize was established, to be awarded in 1822 to the author of "the best work or memoir in pure or applied mathematics which had appeared, or which had been communicated to the Academy, in the space of the two years accorded the contestants." *P.V.* 7, p. 18. (Session of 6 March, 1820). A solid contribution to the resolution of Fermat's problem would have stood a good chance of carrying away the 3,000 francs. Curiously, a receipt for a memoir submitted in competition is found among the papers of Sophie Germain – Bibliothèque Nationale, MS. Fr. 9118 – but information obtained from the Archives of the Academy reveals that this belonged to Libri. In 1856 Fermat's problem was again made the subject of a prize. Once again, no successful resolution was forthcoming, but the 3,000 francs were awarded to Kummer for his investigation of complex numbers composed of roots of unity and of whole numbers.

[4] A friend of Gauss, while visiting Paris, had called on Sophie Germain moving her to make this attempt to renew what had once been an engrossing exchange of letters. There is no evidence, however, that Gauss responded. The letter itself is held at the Universitäts-bibliothek, Göttingen.

[5] Bits and pieces of Sophie Germain's work in relation to Fermat's Last Theorem are found scattered through the Bibliothèque Nationale, MS. Fr. 9114 and 9115. MS. Fr. 9114 ff. 198–207 entitled 'Remarques sur l'impossibité de satisfaire $x^p + y^p = z^p$' is the most

polished draft of her work. Its contents are similar to those in her letter to Gauss.

[6] Legendre, A. M.: 1827, 'Recherches sur quelques objets d'analyse indeterminée et particulièrment sur le théorème de Fermat', *Mém. Acad. Royal des Sci. de l'Institut de France* 6.

[7] Joseph Fourier, (1768–1830), jailed on two occasions during the years of revolution, assistant lecturer in mathematics at the Ecole Polytechnique in 1795, accompanied Napoleon on his ill-fated voyage to Egypt. He was appointed Prefect of the department of Isère in 1801, made a baron in 1808, and, although able to visit Paris only rarely, composed his now-classic treatise on the conduction of heat in solids during this period. In the aftermath of Napoleon's reclamation of power in 1815, he was made Prefect of the Rhône, but for reasons that are not entirely clear, was relieved of his position within a few months. He then moved to Paris in order to devote himself full-time to mathematical pursuits. On his arrival, he might have been able to receive the recognition of several members of the Institute, but he had no job and little money. See Herival, J. W.: 1975, *Joseph Fourier, The Man and the Physicist*, Clarendon Press, Oxford. Also Arago, F.: 1838, 'Eloge historique de Joseph Fourier', *Mém. de l'Acad. des Sciences* 14, lxix–cxxxviii. (English translation in Smyth, W. H., *et al.*: 1857, *Biographies of Distinguished Scientific Men*, Longmans, London, pp.242–286.

[8] Bibliothèque Nationale, MS. Fr. (Nouv. Acq.) 4073. Also in Henry, C.: 1879, 'Les manuscrits de Sophie Germain – documents nouveaux', *Revue phil.* 8, p. 630. While May 2nd fell on a Thursday in 1822 as well as in 1816, the introductory tone of this letter suggests the latter date. The subject of the engraving is unknown – perhaps it was Euler.

[9] Bibliothèque Nationale, MS. Fr. (Nouv. Acq.) 4073. Also in Henry, C., *op. cit.*, p. 629.

[10] Arago, F.: 1838, *op. cit.*, lxxii.

[11] Fourier prevailed, however. In 1812, his research earned him the First Class prize in mathematics. Laplace's respect for his work was an important element in the establishment, as well as award, of this prize. Fourier was elected to the Academy of Sciences in 1817.

[12] Bibliothèque Nationale, MS. Fr. 9118. Also in Stupuy, *op. cit.*, p. 318.

[13] *Ibid.*, p. 323.

[14] A royal ordinance dated 21 March, 1816, occasioned a reorganization of the Institute into four academies – the French Academy, the Royal Academy of Inscriptions and Belles Lettres, the Royal Academy of Sciences, and the Royal Academy of Beaux Arts. *P.V.*, vol. 6, p. 40. (Extraordinary session of 27 March, 1816.)

[15] Bibliothèque Nationale, MS. Fr. 9118. Also in Stupuy, *op. cit.*, p. 324.

[16] Bibliothèque Nationale, MS. Fr. (Nouv. Acq.) 4073. Also appearing in Henry, C., *op. cit.*, p. 627, but with the year of this letter given as 1825. Clearly this is a misprint; in MS. 4073 it appears as 1821 and, furthermore, Delambre had died in 1822.

[17] Germain, S.: 1821, *Reserches sur la théorie des surfaces élastiques*, Mme. V. Courcier, Paris.

[18] Bibliothèque Nationale, MS. Fr. 9118. Also in Stupuy, *op. cit.*, p. 312. While June 1 fell on a Thursday in 1826 as well as in 1820, by 1825 Fourier's penmanship had deteriorated. This letter was written with a steady hand.

[19] Bibliothèque Nationale, MS. Fr. (Nouv. Acq.) 4073. Also in Henry, C., *op. cit.*, p. 628.

[20] *Ibid.*, p. 631. December 6 fell on a Wednesday in 1826 as well as in 1820, but Sophie Germain's mother had died in 1823.

[21] Germain, S.: 1821, *op. cit.*

[22] Sophie Germain simply argued that the elastic force of the plate is the sum of the elastic

force of one beam lying along the axis defining the maximum curvature of the plate and that of another beam lying along the axis defining the minimum curvature.

[23] Germain, S.: 1821, *op. cit.*

[24] *Ibid.*

[25] *Ibid.* Poisson had another notion of what an appropriate, all-powerful hypothesis was, namely that of 'sensible forces at insensible distances.'

[26] Bibliothèque Nationale, MS. Fr. 9118. Also in Stupuy, *op. cit.*, p. 316.

[27] *Ibid.*, p. 317.

[28] *Ibid.*, p. 315.

[29] Stupuy, *op. cit.*, p. 313.

CHAPTER NINE: EMERGENCE OF A THEORY

[1] Fourier, J.: 1818, 'Note relative aux vibrations des surfaces élastiques . . .', *Bull. Soc. Philom. de Paris*, pp. 129–136. Poisson wrote a short note attacking the same problem employing a more traditional method to obtain a solution. He showed that Fourier's solution was contained in his own. Poisson, S. D.: 1818, 'Sur l'intégrale de l'équation relative aux vibrations des plaques élastiques', *Bull. Soc. Philom. de Paris*, pp. 125–128.

[2] A copy of Navier's *Mémoire sur la flexion des plans élastiques* is held at the Archives, Bibliothèque, Ecole Nationale des Ponts et Chaussées, Paris. Navier, at that time, was not a member of the Royal Academy. How widely he distributed his paper is unknown.

[3] *Ibid.* Navier has affirmed the primacy of, and his own preference for, Fourier's study of the response of an infinite plate to an initial disturbance *vis-à-vis* Poisson's work on the same problem. He wholeheartedly accepted Fourier's method and used it to good advantage in his own work. At the close of the introduction to his lithographed paper Navier again rallied behind Fourier, "One should note . . . that the integral solutions have generally been obtained using methods employed for the first time by M. Fourier in his unpublished piece on the theory of heat. The research contained in this memoir offers a new example of the fertility of these methods and their usefulness in applying them to the arts and to the explanation of natural phenomena."

[4] Navier considered an infinitely small plate element which, before deformation, had the shape of a right circular cylinder whose height was equal to the thickness of the plate. After the plate deformed, he assumed this element took the form of a truncated cone, one of its bases expanding and the other contracting. The expansion or contraction was not generally the same in different directions, e.g. as one moved around the axis of the cone. However, expansion and contraction were a maximum (minimum) in the direction corresponding to the direction of maximum (minimum) curvature. All internal forces were assumed to act perpendicular to the faces of the cone. (Here is where Navier's model is deficient. Only in one very special situation – when the cylindrical element deforms into a true cone – is this assumption physically plausible.) Navier then assumed that the intensity of these forces was proportional to their distance from the mid-surface of the plate. With this element and these assumptions, Navier showed that the elastic moment in any direction is proportional to

$$h^3(1/r)\delta(1/r)$$

where h is the plate's thickness, and r the radius of curvature in that particular direction. Integrating over the conical element, Navier obtained the following expression for its elastic moment:

$$\frac{1}{2} \epsilon h^3 \delta \left[\left(\frac{1}{\rho} + \frac{1}{\rho'} \right)^2 - \frac{1}{3\rho\rho'} \right]$$

Here ϵ is a constant that depends on the material from which the plate was made and ρ and ρ' the two principal curvatures. Navier relates how Poisson had discussed the way in which the above expression yields the partial differential equation first obtained by Lagrange (see note 25, Chapter 6). It differs from Sophie Germain's expression for the elastic moment in two ways: first, the cube of the thickness appears in Navier's relationship, whereas Sophie Germain deduced that the fourth power should enter; second, Navier's expression contains a term proportional to the product of the principal curvatures. It was this term that Sophie Germain, in a footnote to her first entry, suggested might enter but could be neglected if the plate was assumed to experience only small deflections (see Figure V.4). Surprisingly, Navier makes the same error – he too claims that this term can be neglected.

[5] In Timoshenko, S.: 1959, *Theory of Plates and Shells*, (2nd ed.), McGraw-Hill, New York. The author, in one section of this modern treatise, has reproduced and embellished Navier's analyses of several phenomena, including the deflection of a uniformly loaded, simply supported, rectangular plate and the deflection of the same plate subjected to a concentrated load.

[6] See Timoshenko, S.: 1953, *History of the Strength of Materials*, McGraw-Hill, New York. Todhunter, I., and Pearson, K.: 1886, *A History of the Theory of Elasticity and of the Strength of Materials*, At the University Press, Cambridge. An extract of Navier's memoir was published in the *Bull. Soc. Philom. de Paris* (1823), pp. 177–181. His complete analysis was not published until 1827. Navier's approach yields a one-constant theory of elasticity. We know today that two independent constants are required to specify the material properties of an isotropic, elastic solid. The question of how many independent constants were required to define the character of an elastic body was addressed by a variety of nineteenth-century scientists including Green, Lamé, Stokes, Clausius, Kirchoff, and St. Venant, as well as Cauchy and Poisson.

[7] *P.V.*, vol. 7, p. 296. An extract of Navier's analysis of the motion of fluids appeared in the *Bull. Soc. Philom. de Paris* (1825): pp. 49–52. The complete memoir is found in the *Mém. Acad. Sci.* 6, pp. 389–440.

[8] *P.V.*, vol. 7, p. 370.

[9] *Ibid.*, p. 371.

[10] Cauchy, 1823, 'Recherches sur l'équilibre et le mouvement intérieur des corps solides ou fluides, élastiques ou non-élastiques', *Bull. Soc. Philom. de Paris*, pp. 9–13. At the end of this abstract Cauchy promised, in a future memoir, to analyze the behavior of elastic plates and beams. As noted in the text, Navier's memoir on plates to which Cauchy refers – "a memoir on plates published the 14th August 1820 . . ." – had appeared only in the form of a privately distributed lithograph edition.

[11] Cauchy toyed with two types of possible stress-strain relations. First he assumed that the three principal stresses at a point were respectively proportional to the three principal strains, e.g. $\sigma_I = k\epsilon_I$, $\sigma_{II} = k\epsilon_{II}$, $\sigma_{III} = k\epsilon_{III}$. This yields a one-constant, (k), theory of elasticity similar to, but *not* the same as, that derived by Navier. Second, in a modification of this hypothesis, he assumed stress depended upon strain in the following way

$$\sigma_I = k\epsilon_I + Kv, \sigma_{II} = k\epsilon_{II} + Kv, \sigma_{III} = k\epsilon_{III} + Kv$$

where *K is a second elastic constant and* ^{v}, the volume dilatation. He noted that setting $K = k/2$ yields Navier's equations of equilibrium in terms of displacement. The details of Cauchy's analysis appear in Cauchy, A.: 1828, *Exercises de mathématiques, année 1828*, in vol. 8 of *Oeuvres Complètes*, Sér. 2, 13 vols., Gauther-Villars, Paris, 1897–1958. There, in a section 'Sur les équations qui expriment les conditions d'équilibre ou les lois du mouvement intérieur d'un corps solide, élastique ou non-élastique', Cauchy states that most of the equations derived in this section had been drawn from his memoir read to the Academy on 30 September, 1822, (p. 215). For an evaluation of Cauchy's work see Truesdell, C.: 1961, 'Stages in the Development of the Concept of Stress', in *Problems of Continuum Mechanics*, Muskhelishvili Anniversary Volume, (English edn.), Soc. for Indust. and Appl. Math., Philadelphia. See also by the same author, Truesdell, C.: 1962, 'The Creation and Unfolding of the Concept of Stress', *Essays in the History of Mechanics*, Springer-Verlag, Berlin.

[12] Navier, C.: 1823, 'Observations communiquées par M. Navier, à l'occasion du Mémoire de M. Cauchy," *Bull. Soc. Philom. de Paris*, pp. 36–37.

[13] Navier, C.: 1823, 'Extrait des recherches sur la flexion des plans élastiques', *Bull. Soc. Philom. de Paris*, pp. 92–102. Navier's description of how he derived the plate equation in this abstract of his work differs significantly from his description in his lithographed, 1820 memoire. The extract reads as though Navier had deduced his expression for the elastic moment from a molecular model. In his 1820 memoir he worked without making any reference to sensible forces at insensible distances. A summary of Navier's 'Sur les lois de l'équilibre et du mouvement des corps solides élastiques' was published in *Bull. Soc. Philom. de Paris* (1823) 177–181.

[14] *P.V.*, vol. 8, p. 9 (Session of 19 January, 1824).

[15] Letter from Fourier to Sophie Germain, undated but obviously written shortly after Navier had read his memoir to the Academy, held at the Bibliotèque Nationale, MS. Fr. (Nouv. Acq.) 4073. Also in Henry, C.: 1879, 'Les manuscrits de Sophie Germain', *Revue Phil.* 8, p. 630.

[16] Navier, 'Extrait', *op. cit.*, p. 93.

[17] Poisson deduced that thickness should enter as h^2, Navier showed that it should appear as h^3, and Sophie Germain derived h^4. Navier's result is correct and cogently argued.

[18] Letter from Fourier to Sophie Germain dated 15 March, 1824. Bibliothèque Nationale, MS. Fr. 9118. Also in Stupuy, *op. cit.*, p. 320. The postscript indicates that this letter was written the Friday after the presentation of her memoir.

[20] The appearance in print of Germain, S.: 1880, *Mémoire sur l'emploi de l'épaisseur dans la théorie des surfaces élastiques*, Gauthier-Villars, Paris, followed Stupuy's publication of her *Oeuvres philosophiques* in 1879. The cover page of Sophie Germain's manuscript, which until 1879 had been closeted away with Prony's papers in the Archives of the Ecole des Ponts et des Chaussées, indicates that Poisson alone had perused her work. Her manuscript is now held at the Archives of the Academy of Sciences, Paris.

[21] Germain, S.: 1826, *Remarques sur la nature, les bornes et l'étendue de la question des surfaces élastiques*, Huzard-Courcier, Paris. Here Sophie Germain provides a geometrical interpretation of her expression for the elastic force of a plate; this force is simply measured by the radius of curvature of an equivalent sphere, the radius of this sphere being equal to the mean curvature at a point in the deformed plate. In her introduction she noted that her previous memoir had been presented to the Academy two years before but Prony and Poisson had not, as yet, made their report. She promised to publish this earlier treatise when

circumstances allowed.

[22] Letter from Fourier to Sophie Germain dated July 24, 1826. Bibliothèque Nationale, MS. Fr. 9118. Also in Stupuy, *op. cit.*, p. 327, but with the year in error.

[23] Draft of a letter from Sophie Germain lacking address and dated 18 July. Bibliothèque Nationale, MS. Fr. 9118. Also in Stupuy, *op. cit.*, p. 328. The draft's contents reveal that the letter was intended for Cauchy. The date of Cauchy's response establishes, in turn, the year it was written.

[24] Cauchy's memoirs in elasticity were not generally accessible until the period 1827–1829, when he published them in his *Exercises de mathématiques*. See his *Oeuvres complètes, op. cit.*, 7, 8, and 9.

[25] Bibliothèque Nationale, MS. Fr. 9118. Also in Stupuy, *op. cit.*, p. 326, but with the year in error.

[26] At the Academy's session of 1 October, 1827, Poisson read a short note describing his research in progress on the vibration of solids. He asked the Academy's permission to make known the basis of his analysis and some of its results even though he had not finished his work. At the conclusion of his reading, Cauchy announced that he too had been working for a long while on this subject – the equilibrium and movement of a solid body considered as a system of isolated molecules. At the end of the meeting Cauchy delivered a manuscript for review. (*P.V.*, vol. 8, p. 603.) Poisson's note was published in the *Bull. des Sci. Math.*, (ed. Saigney), 9, (1828) 27–31. Cauchy's molecular analysis was published as 'Sur l'équilibre et le mouvement d'un système de points matériels sollicités par les forces d'attraction ou de repulsion mutuelle', *Exercices de mathématiques*, année 1828. *Oeuvres complètes, op. cit.* 8, pp. 227–252. Navier's complete memoir appears in *Mém. Acad. Sci.* 7, pp. 375–393.

[27] Poisson, S. D.: 1828, 'Mémoire sur l'équilibre et le mouvement des corps élastiques', *Annal. de chimie* 37, pp. 337–355. His full treatise, carrying the same title, is in *Mém. Acad. Sci.* 8, pp. 357–570, 623–627.

[28] Navier, C.: 1828, 'Note relative à l'article intitulé: Mémoire sur l'équilibre', *Annal. de chimie* 38, pp. 304–314.

[29] Poisson responded with 'Réponse à une note de M. Navier . . .', *Annal. de chimie* 38, (1828), 435–440. Navier, in turn, submitted his 'Remarques sur l'article de M. Poisson, inséré dans le cahier d'août, page 435, *Annal. de chimie* 39, (1828) 145–151. The polemic terminated with a letter from each participant to Arago, a convenient third party: 'Lettre de M. Poisson à M. Arago', *Annal. de chimie* 39, (1828) 204–211 and 'Lettre de M. Navier à M. Arago', *Annal. de chimie* 40, (1829) 99–107.

[30] Navier was correct on this point.

[31] This note now can be found among Sophie Germain's paper. Bibliothèque Nationale, MS. Fr. 9118.

[32] Germain, S.: 1828, 'Examen des principes qui peuvent conduire à la connaissance des lois de l'équilibre et du mouvement des solides élastiques', *Annal. de chimie* 38, pp. 123–131.

[33] *Ibid.*

CHAPTER TEN: FINAL YEARS

[1] Lherbette, (ed.): 1833, *Considérations sur l'état des sciences et lettres*, Lachevardière, Paris.

[2] Bibliothèque Nationale, MS. Fr. 9114. Also in Stupuy, *op.*

[3] Stupuy's first edition was published in 1879.

[4] Germain, S.: 1831, 'Memoire sur la courbure des surfaces', *Jour. für die reine und angewandte Mathematik*, (ed. A. L. Crelle), 7, pp. 1–29. Germain, S.: 1831, 'Note sur la manière dont se composent les valeurs de y et z dans l'équation $4(x^p - 1)/(x - 1) = y^2 \pm pz^2 \ldots$' *ibid.*, pp. 201–204.

[5] See Sophie Germain's letter to Cauchy and note 21 in chapter 9.

[6] Gauss, D. F.: 1827, *Disquisitiones generales circa superficies curvas*, Göttingen. A French translation was published in 1852, *Nouvelles annales de mathématiques* 11, pp. 195–252.

[7] Letter from Sophie Germain to Gauss dated 28 March, 1829, held at the Universitäts-bibliothek, Göttingen.

[8] Libri, G.: 1832, 'Notice sur Mlle. Sophie Germain', in Lherbette, *op. cit.*

[9] *P.V.*, vol. 7, p. 131 (Session of 22 January, 1821). Note of Cauchy's report is found in the minutes of the Academy's next session.

[10] In a letter from Florence dated 17 November, 1825, Libri asked Sophie Germain to chastise Fourier, in order to move the Perpetual Secretary to speed up the Academy's review of a memoir on number theory he had submitted six months earlier. In this same letter Libri informed Sophie Germain that he had sent her a memoir by Riccati that she wanted. He hoped that this would ". . . help you to recall one who has sentiments of profound esteem for you. May I hope to receive news from you from time to time? I hope that you, being so kind, will agree to my request, thinking of the pleasure it will being me. Speak of your work, your amusements, music, and of everything that interests you for that will interest me as well." Bibliothèque Nationale, MS. Fr. (Nouv. Acq.) 4073. Also in Henry, C.: 1879, 'Les manuscrits de Sophie Germain – documents nouveaux', *Revue phil.* 8, p. 636. Another letter bound in this same collection, from Fourier to Sophie Germain, indicates that she fulfilled Libri's request.

[11] Libri, *op. cit.*

[12] See note 25, Chapter 6.

[13] Letter from Sophie Germain to Gauss dated 22 May, 1809, held at the Universitäts-bibliothek, Göttingen.

[14] Bibliothèque Nationale, MS. Fr. (Nouv. Acq.) 4073. Also in Henry, C.: 1879, *op. cit.*, p. 631.

INDEX

STUDIES IN THE HISTORY
OF MODERN SCIENCE

Editors:

ROBERT S. COHEN (Boston University)
ERWIN N. HIEBERT (Harvard University)
EVERETT I. MENDELSOHN (Harvard University)